A Study Guide to Chemical Pathology:

A Progressive, Three-level Approach

A Study Guide to Chemical Pathology:
A Progressive, Three-level Approach

M. DESMOND BURKE, MD
Associate Professor
Department of Laboratory Medicine and Pathology
University of Minnesota
and
Associate Pathologist
Mt Sinai Hospital
Minneapolis

ROBERT C. ROCK, MD
Director
Department of Laboratory Medicine
The Johns Hopkins Hospital
Baltimore

Educational Products Division
American Society of Clinical Pathologists
Chicago

ISBN 0-89189-075-0

This complimentary copy of A STUDY GUIDE TO CHEMICAL PATHOLOGY: A PROGRESSIVE, THREE-LEVEL APPROACH by M. Desmond Burke, MD, and Robert C. Rock, MD, is sent to you by the American Society of Clinical Pathologists (ASCP). The book can be used for self-teaching or as a guide to developing a formal course in Clinical Chemistry. We hope it will be of use to you. To purchase additional copies, write to:

ASCP Customer Services Dept.
PO Box 12073
Chicago, IL 60612

CONTENTS

PREFACE

For more than 50 years, the American Society of Clinical Pathologists has devoted considerable effort to "continuing" the education of its members. The current range, scope, and high quality of the Society's workshops and educational materials attest to the success of that effort. To guarantee ongoing success, the continuing medical education process itself must constantly be subject to reevaluation and improvement—in structure as well as in scope. Questions should be raised and efforts should be made to pinpoint and meet the educational needs of students and professionals in pathology and in other areas of health-care. ASCP must determine the reasons for a member's current choice of workshops, seminars, and educational materials in order to develop and offer new programs and publications that will satisfy future needs. Above all, to continue to be an effective educator, ASCP must be ever-aware of its role in developing, promoting, and advancing medical knowledge.

In 1977, these considerations led ASCP's Commission on Continuing Education Council on Undergraduate and Graduate Education to recommend a more structured learning curriculum. This recommendation, in turn, led to the formation of A STUDY GUIDE TO CHEMICAL PATHOLOGY: A PROGRESSIVE, THREE-LEVEL APPROACH.

The Guide's major divisions, "Technical Methods," "Chemical Pathology," "Diagnostic Application of Laboratory Data," and "Laboratory Administration," derive as much from the legacy of early 19th century patho-clinicians and the testament of Virchow's Cellular Pathology as they are a result of our historic self-charged responsibility to modern science.

The purpose of the Guide's three-tiered, content-goals-learning aids approach is twofold: to aid the "student" in developing a self-instructional curriculum, and to assist ASCP in structuring educational programs that meet the demands of modern pathology.

Admittedly, many of the goals suggested at the highest training levels are ones to which we can only aspire. Nevertheless, it *is* our responsibility to try. Science is, above all, an explanatory but unfinished endeavor. The term *pathology* still means "the study of disease" and by the choice of goals set forth in this Study Guide we take that definition literally. In our view, it is as much of a mistake to confuse science with technology as it is to think of clinical pursuits as unscientific—and we frequently are guilty of both.

We hope that this structured blend of technical methods, chemical pathology, diagnostic application of laboratory data, and laboratory management will serve to remind pathologists of their classical heritage. If our work has contributed to a revival of pathology's traditional "watchdog" role we are pleased with the accomplishment.

Alvan Feinstein's views are pertinent here. Having avowed that modern pathologists err in confining themselves unduly to morphologic pursuits, he says ". . . If the clinician of the 19th century could travel to the necropsy room to learn about 'disease,' the pathologist of the 20th century may need to revise the direction of migration in order to maintain his central coordinating role in the study of 'disease'."*

MDB
RCR

*Feinstein AR: Clinical Judgment. Baltimore: Williams & Wilkins Co, 1967, p 108

INTRODUCTION

This Study Guide is designed to offer readers assistance in structuring their learning in clinical chemistry. In this work, clinical chemistry is broadly divided into four areas: technical methods, chemical pathology, diagnostic application of laboratory data, and laboratory administration. Each major area is further subdivided as the authors deemed appropriate. The taxonomy used throughout is the result of a compromise, in that some overlap in topics is inevitable. Nevertheless, we believe that the material outlined is inclusive, and that it encompasses most aspects of hospital chemical pathology.

Topics are subdivided into two or three levels of performance: Levels I, II, and III. This substructure was devised by one of us (MDB) and is arbitrary to a large extent. We make the following suggestions regarding the reader's level of performance: Practicing pathologists without a particular interest or responsibility in chemical pathology should attempt to attain Level I proficiency in the application of the technical methods and clinical entities commonly encountered in their practices. Practicing pathologists with a particular interest or responsibility in clinical chemistry should try to achieve a more comprehensive Level I proficiency overall, and a Level II proficiency in the application of technical methods and aspects of chemical pathology directly relating to their practices.

Physicians whose practices are devoted to clinical chemistry should attempt to achieve an overall, comprehensive Level II proficiency, and should aspire to Level III proficiency in areas of specific responsibility or interest. Laboratory scientists who devote all of their time to clinical chemistry should achieve comprehensive Level II proficiency in technical methods, and should attempt Level III proficiency in areas of special interest or responsibility. Furthermore, laboratory scientists should attempt Level I proficiency in diagnostic application of laboratory data. Senior technologists might be expected to achieve Level I proficiency in technical methods.

Residents in pathology seeking to practice general pathology should attempt to achieve Level I proficiency in all areas. Those residents considering a career in clinical chemistry should try to reach an overall, comprehensive Level II proficiency.

Each topic is subdivided to include goals and learning aids. No doubt that stating measurable behavioral objectives would be preferable; nevertheless, we are, in a way, charting new ground and we feel that measurable objectives could, at this time, be stated in general terms only. For example, for the pathologists with primary responsibility in chemical pathology, a reasonable behavioral objective would be sufficient familiarity with the subject to influence test ordering habits and provoke consultations. As of now, this behavior probably is not the rule.

Learning aids cited throughout this publication include a list of general works (Suggested References at the end of this Study Guide), bibliography, and assorted ASCP teaching aids: texts and atlases, audiovisual aids, and a miscellaneous, unclassified group of educational materials. Learning aids distributed by other societies and groups, and those offered at a variety of ASCP-sponsored national and regional meetings, seminars, workshops, demonstrations, and exhibits are not cited in this work.

ASCP learning aids are notably lacking in many areas of chemical pathology (as opposed to technical methods). We hope that this publication, in one way or another, will serve to offset that lack.

Virtually all of the references cited in the technical methods and laboratory management divisions were provided by one of us (RCR), and the majority of references pertaining to chemical pathology and diagnostic application were the responsibility of the other (MDB). With few exceptions, the references cited have a distinctly personal quality in that they are found in the files of the respective author.

With this study guide we hope to achieve the following: (1) Given a content outline, the reader will be able to appreciate the scope of clinical chemistry; (2) Given the content outline's three-level structure, the reader will be able to identify content designated at Levels I, II, or III; (3) Given goals for each content topic at Levels I, II, or III, the reader will better appreciate the range of proficiency required to play a more meaningful role as a chemical pathology consultant; and (4) Given a list of references and appropriate ASCP teaching aids for each content topic at Levels I, II, or III, the reader will make use of the educational materials necessary to structure a curriculum in clinical chemistry.

Section 1

TECHNICAL METHODS

GENERAL PROCEDURES

LEVEL I

Content

Patient preparation, specimen collection and preparation, reagent preparation, standards, units of measurement, basic laboratory mathematics

Quality control

Laboratory safety

Goals

The ability to ensure high standards of quantitative measurement

Command of required laboratory mathematics, including the use of logarithms, understanding of exponents and significant figures, ability to express and calculate dilutions, and understanding of the SI unit system

Understanding of basic statistical theory and the ability to perform statistical tests (standard deviation, coefficient of variation, correlation *t* tests, F tests, linear regression analysis). Ability to prepare control preparations, to participate in interlaboratory calibration of spectrophotometers, to monitor laboratory reagents and chemicals, to ensure preventive maintenance of instruments, and to follow strategy in developing new methods

Maintenance of laboratory safety, including awareness and avoidance of the potential dangers of chemical, electrical, and biological hazards; ability to handle and store dangerous chemicals, to dispose of radioactive wastes, and to control infection; and understanding of proper grounding of instruments

Learning Aids

Bibliography

Annual Reports on College of American Pathologists Surveys. Am J Clin Pathol 54(suppl):435–530, 1970

Bermes EW, Erviti V, Forman DT: Statistics, normal values, and quality control, in Tietz NW (ed): Fundamentals of Clinical Chemistry, ed 2. Philadelphia: WB Saunders Co, 1976, p 60–102

Bermes EW, Forman DT: Basic laboratory principles and procedures, in Tietz NW (ed): Fundamentals of Clinical Chemistry, ed 2. Philadelphia: WB Saunders Co, 1976, p 1–59

Boutwell JH: A National Understanding for the Development of Reference Materials and Methods for Clinical Chemistry: Proceedings of a Conference. American Association of Clinical Chemistry. Atlanta, Nov 16–17, 1977

Percy-Robb IW, Proffitt J, Whitby LG: Precautions adopted in a clinical chemistry laboratory as a result of an outbreak of serum hepatitis affecting hospital staff. J Clin Pathol 23:751–756, 1970

Pragay DA: Pollution control and suggested disposal guidelines for clinical chemistry laboratories. Clin Chem 21:1839–1844, 1975

Provisional Recommendation on Quality Control in Clinical Chemistry. International Federation of Clinical Chemistry: Committee on Standards. A six-part series:
(1) General principles and terminology. Clin Chem 22:532–540, 1976
(2) Assessment of analytical methods for routine use. Clin Chem 22:1922–1932, 1976
(3) Calibration and control materials. Clin Chem 23:1784–1789, 1977
(5) External quality control. Clin Chem 24:1213–1220, 1978
(6) Quality requirements from the point of view of health care. Clin Chem 23:1066–1071, 1977

Radin N: What is a standard? Clin Chem 13:55–76, 1967

Saracci R: The power (sensitivity) of quality-control plans in clinical chemistry. Am J Clin Pathol 62:398–406, 1974

Thiers RE, et al: Sample stability: A suggested definition and method of determination. Clin Chem 22:176–183, 1976

Westgard JO, de Vos DJ, Hunt MR: Concepts and practices in the evaluation of laboratory methods. I. Background and approach. II. Experimental procedures. III. Statistics. IV. Decisions on acceptability. V. Applications. Am J Med Technol 44:290–300, 420–430, 552–571, 727–742, 803–813, 1978

Westgard JO, Hunt MR: Use and interpretation of common statistical tests in method-comparison studies. Clin Chem 19:49–57, 1973

Whitehead TP: Quality Control in Clinical Chemistry. New York: John Wiley & Sons Inc, 1977

Wu GT, Twomey SL, Thiers RE: Statistical evaluation of method-comparison data. Clin Chem 21:315–320, 1975

Texts (ASCP)

Council on Clinical Chemistry

Barnett RN (ed): Statistical Methods in the Clinical Laboratory. Chicago: American Society of Clinical Pathologists, 1968 (Cat. no. 45-2-016-00—out of print)

Beeler MF: Interpretations in Clinical Chemistry. Chicago: American Society of Clinical Pathologists, 1978 (Cat. no. 45-2-036-00)

Copeland BE: Quality Control in Clinical Chemistry. Chicago: American Society of Clinical Pathologists, 1973 (Cat. no. 45-2-013-00—out of print)

Copeland BE: Quality Control in Clinical Chemistry Kit. Chicago: American Society of Clinical Pathologists, 1973 (Cat. no. 45-2-014-00—out of print)

Other Councils

Marymont JH Jr: Laboratory Mathematics. Chicago: American Society of Clinical Pathologists, 1973 (Cat. no. 45-9-003-00—out of print)

Matheson LV (ed): An Introduction to the Role of the Computer in the Laboratory—Basic Fundamentals for Medical Laboratory Personnel. Chicago: American Society of Clinical Pathologists, 1973 (Cat. no. 45-0-001-00—out of print)

Mukherjee KL: Introductory Mathematics for the Clinical Laboratory.

Chicago: American Society of Clinical Pathologists, 1978 (Cat. no. 45-9-006-00)

Prince JR, Schmidt LD: Statistics and Mathematics in the Nuclear Medicine Laboratory. Chicago: American Society of Clinical Pathologists, 1976 (Cat. no. 45-8-007-00)

Tischer RG, Turner BH, Hogan SC: Problem Solving in Medical Technology and Microbiology. Chicago: American Society of Clinical Pathologists, 1979 (Cat. no. 45-7-010-00)

Audiovisual Aids (ASCP)

Copeland BE: What Is Quality Control? Audiotape 44. Chicago: American Society of Clinical Pathologists (Cat. no. 18-0-044-06–out of print)

Hoffman G: Programmable Desktop Calculators. Audiotape 35. Chicago: American Society of Clinical Pathologists (Cat. no. 18-0-035-06–out of print)

Miscellaneous Learning Aids (ASCP)

Beeler MF: Going metric, in Baer DM (ed): Technical Improvement Service. Number 16. Shilling JM (section ed). Chicago: American Society of Clinical Pathologists, 1974, pp 8–21 (Cat. no. 63-0-01-00–out of print)

Duckworth JK, Steven MV, Hamlin WV: Preventive instrument maintenance in the clinical laboratory, in Baer DM (ed): Technical Improvement Service. Number 17. Shilling JM (section ed). Chicago: American Society of Clinical Pathologists, 1974, pp 7–16 (Cat. no. 63-0-017-00–out of print)

Free AH, Free HM: The collection and preservation of urine, in Eggen RR (ed): Technical Improvement Service. Number 13. Chicago: American Society of Clinical Pathologists, 1973, pp 67–85 (Cat. no. 63-0-013-00–out of print)

Nichols AH: How to use the reference laboratory. Part 1: The make/buy decision, in Eggen RR (ed): Technical Improvement Service. Number 13. Chicago: American Society of Clinical Pathologists, 1973, pp 59–66 (Cat. no. 63-0-013-00–out of print)

Tenczar FJ, Kowalski TA: The centrifuge: Maintenance and calibration, in Baer DM (ed): Technical Improvement Service. Number 14. Chicago: American Society of Clinical Pathologists, 1973, pp 72–80 (Cat. no. 63-0-013-00–out of print)

INSTRUMENTATION: PRINCIPLES AND EQUIPMENT

LEVEL I

Content

General laboratory equipment: analytic balance, volumetric equipment

Instrumentation: basic electrical and electronic principles

Spectrophotometry: visual and ultraviolet

Turbidimetry and nephelometry

Flame photometry

Electrophoresis

Chromatography: plane, ion exchange, gel filtration

Osmometry

Continuous flow automation

Discrete sample automation

Electrochemical measurement: potentiometry

Radiochemical measurement

Goals

The ability to solve chemical and instrumental problems through an understanding of underlying principles, reinforced by workshop-type problem-solving sessions

Familiarity with volumetric equipment and the analytic balance

Prerequisite knowledge of basic electrical and electronic theory, the terminology and practical aspects of circuitry, and the principles of detector mechanisms for each instrumental method of analysis

Comprehension of the underlying theoretic, physical, or chemical principles of each instrumental method of analysis to ensure intelligent application:

principles of absorption and emission of radiation; electronic, vibrational, or rotational energy changes for photometry, spectrophotometry, electrophoresis, chromatography, and electrochemical and radiochemical instrumental measurement

understanding of the properties of liquids and of phase diagrams for osmometry

theoretic principles of continuous flow automation and of flow dynamics for the practical aspects of autoanalyzer measurement

Learning Aids

Bibliography

General

Bender GT: Chemical Instrumentation: A Laboratory Manual on Clinical Chemistry. Philadelphia: WB Saunders Co, 1972

Hicks R, Schencken JR, Steinrauf MA: Laboratory Instrumentation. New York: Harper & Row Publishers Inc, 1974

White WL, Erickson MM, Stevens SC: Practical Automation for the Clinical Laboratory, ed 2. St Louis: CV Mosby Co, 1972

Spectrophotometry

Rand RN: Practical spectrophotometric standards. Clin Chem 15: 839–863, 1969

Flame Photometry

Rains TC, Dean JA (eds): Flame Emission and Atomic Absorption Spectrometry. New York: Marcel Dekker Inc, 1971

Electrophoresis

Cawley LP: Electrophoresis and Immunoelectrophoresis. Boston: Little Brown & Co, 1969

Chromatography

Dean JA: Chemical Separation Methods. New York: Van Nostrand Reinhold Co, 1969

Continuous-flow Automation

Thiers RE, Cole RR, Kirsch WJ: Kinetic parameters of continuous flow analysis. Clin Chem 13:451–467, 1967

Thiers RE, Oglesby KM: The precision, accuracy, and inherent errors of automatic continuous flow methods. Clin Chem 10:246–257, 1964

Walker WH, Pennock CA, McGowan GK: Practical considerations in kinetics of continuous-flow analysis. Clin Chem Acta 27:421–435, 1970

Discrete Sample Automation

Broughton PM, et al: A revised scheme for the evaluation of automatic instruments for use in clinical chemistry. Ann Clin Biochem 11: 207–218, 1974

Moss DW: Automatic enzyme analyzers. Adv Clin Chem 19:1–56, 1977

Electrochemical Measurement

Durst RA (ed): Ion-Selective Electrodes. Special publication no. 314. Washington, DC: US Government Printing Office, 1969

Radiochemical Measurement

Horrocks DL: Standardizing ^{125}I sources and determining ^{125}I counting efficiencies of well-type gamma counting systems. Clin Chem 21:370–375, 1975

Texts and Atlases (ASCP)

Council on Clinical Chemistry

Cawley LP, et al: Basic Electrophoresis, Immunoelectrophoresis and Immunochemistry. Chicago: American Society of Clinical Pathologists, 1972 (Cat. no. 45-2-030-00—out of print)

Cawley LP, Minard BJ, Penn GM: Electrophoresis and Immunochemical Reactions in Gels: Techniques and Interpretation. Chicago: American Society of Clinical Pathologists, 1978 (Cat. no. 45-2-035-00)

Fleischer WR, Gambino SR: Blood pH Po_2, and Oxygen Saturation. Chicago: American Society of Clinical Pathologists, 1972 (Cat. no. 45-2-029-00 – out of print)

Other Councils

Howard PL: Basic Liquid Scintillation Counting. Chicago: American Society of Clinical Pathologists, 1976 (Cat. no. 45-8-006-00)

Penn GM, Batya J: Interpretation of Immunoelectrophoretic Patterns. Chicago: American Society of Clinical Pathologists, 1978 (Atlas and slides: Cat. no. 15-A-001-00/Atlas only: Cat. no. 16-A-001-00)

Prince JR, Schmidt LD: Statistics and Mathematics in the Nuclear Medicine Laboratory. Chicago: American Society of Clinical Pathologists, 1976 (Cat. no. 45-8-007-00)

Trainer TD, Howard PL, Hodnett JP: Radioisotopes in the Clinical Laboratory. Chicago: American Society of Clinical Pathologists, 1972 (Atlas and slides: Cat. no. 15-8-001-00/Atlas with microfiche: Cat. no. 17-8-001-00)

Audiovisual Aids (ASCP)

Fitzsimons E: Standards for Automated Multichannel Analyzers. Audiotape 36. Chicago: American Society of Clinical Pathologists, (Cat. no. 18-0-036-06 – out of print)

Fleischer WR, Hartmann AE: Blood Gases and Their Measurement – Part I. Chicago: American Society of Clinical Pathologists, 1976 (Cat. no. 47-2-022-00)

Fleischer WR, Hartmann AE: Blood Gases and Their Measurement – Part II. Chicago: American Society of Clinical Pathologists, 1976 (Cat. no. 47-2-024-00)

Miscellaneous Learning Aids (ASCP)

Gambino SR (ed): Osmolality, in Selected Clinical Chemistry "Check Sample" Critiques, vol 2. Chicago: American Society of Clinical Pathologists, 1975 (Cat. no. 45-2-032-00 – out of print)

LEVEL II

Content

Atomic absorption spectrophotometry

Fluorometry

Chromatography: theoretical principles; gas and high-pressure liquid chromatography

Centrifugal analyzers

Electrochemical measurement: coulometry, amperometry, voltammetry

Goals

Comprehension of the underlying principles of gas chromatography, electroanalytic techniques, and of instrumental methods and equipment that, in some instances, are more complicated and used less frequently

Learning Aids

Bibliography

Atomic Absorption Spectrometry
Rains TC, Dean JA (eds): Flame Emission and Atomic Absorption Spectrometry. New York: Marcel Dekker Inc, 1971

Fluorometry
Elevitch FR: Fluorometric Techniques in Clinical Chemistry. Boston: Little Brown & Co, 1973

Chromatography
Dixon PF, Stoll MS: High-pressure liquid chromatography in clinical chemistry. Ann Clin Biochem 13:409–432, 1976

McNair HM, Bonelli EJ: Basic Gas Chromatography, ed 5. Berkeley, Calif: Varian Aerograph, 1969

Symposium: Gas chromatography in clinical chemistry. Clin Chim Acta 34:131–365, 1972

Centrifugal Analyzers
Hatcher DW, Anderson NG: GeMSAEC: A new analytical tool for clinical chemistry total serum protein with biuret reaction. Am J Clin Pathol 52:645–651, 1969

Scott CD, Burtis CA: A miniature fast analyzer system. Anal Chem 45: 327A–340A, 1973

Tiffany TO, et al: Fluorometric fast analyzer: Some applications to fluorescence measurements in clinical chemistry. Clin Chem 19: 871–882, 1973

Electrochemical Measurement

Willard HH, Merritt LL, Dean JA: Instrumental Methods of Analysis, ed 5. New York: D Van Nostrand Co, 1974

Texts (ASCP)

Council on Clinical Chemistry

Fuller JB: Atomic absorption spectrophotometry (AAS), in Fuller JB (ed): Selected Topics in Clinical Chemistry II. Chicago: American Society of Clinical Pathologists, 1975, pp 47–71 (Cat. no. 45-2-033-00—out of print)

Zettner A, Berman E: Clinical Methods of Atomic Absorption Spectroscopy. Chicago: American Society of Clinical Pathologists, 1965 (Cat. no. 45-2-001-00—out of print)

Other Councils

Howard PL: Basic Liquid Scintillation Counting. Chicago: American Society of Clinical Pathologists, 1976 (Cat. no. 45-8-006-00)

Trainer TD, Howard PL, Hodnett JP: Radioisotopes in the Clinical Laboratory. Chicago: American Society of Clinical Pathologists, 1976 (Atlas with Slides: Cat. no. 15-8-001-00/Atlas with microfiche: Cat. no. 17-8-001-00)

Audiovisual Aids (ASCP)

Baer D: GEMSAEC Analyzers. Audiotape 33. Chicago: American Society of Clinical Pathologists, (Cat. no. 18-0-033-06—out of print)

Cawley LP: Principles of Radioimmunoassay—Part I. Chicago: American Society of Clinical Pathologists, 1975 (Cat. no. 47-8-009-00)

Cawley LP: Principles of Radioimmunoassay—Part II. Chicago: American Society of Clinical Pathologists, 1975 (Cat. no. 47-8-010-00)

LEVEL III

Content

Mass spectrometry

Instrumental methods used in the basic sciences as applied to clinical chemistry, including infrared and Raman spectroscopy, nuclear magnetic resonance spectroscopy, electron spin resonance spectroscopy, polarimetry, circular dichroism and optical rotatory dispersion, thermal analysis

Goals

Mastery of complex instrumental techniques and understanding of underlying principles

Learning Aids

Bibliography

Mass Spectrometry

Lawson AM: The scope of mass spectrometry in clinical chemistry. Clin Chem 21:803–824, 1975

Melville RS, Dobson VF: Selected Approaches to Gas Chromatography—Mass Spectrometry in Laboratory Medicine. DHEW Publication No. (NIH) 75-762. Washington, DC: US Government Printing Office, 1975

Roboz J: Mass spectrometry in clinical chemistry. Adv Clin Chem 17: 109–191, 1975

Other Instrumental Approaches

Malmstadt HV, Enke CG: Electronics for Scientists: Principles and Experiments for Those Who Use Instruments. New York: WA Benjamin Inc, 1963

Willard HH, Merritt LL, Dean JA: Instrumental Methods of Analysis, ed 5. New York: D Van Nostrand Co, 1974

ANALYTIC METHODS BY CHEMICAL GROUP

LEVEL I

Content

Carbohydrates—derivatives and metabolites; glucose, ketones, xylose

Lipids and lipoproteins – cholesterol, triglycerides

Proteins – albumin, globulins, fibrinogen, haptoglobin, transferrin, immunoglobulins

Enzymes – lactic dehydrogenase (LDH), creatine phosphokinase (CPK), aspartate aminotransferase (AST), alanine aminotransferase (ALT), amylase, lipase, alkaline phosphatase (ALP), acid phosphatase (ACP), gamma-glutamyl transpeptidase (GGT)

Electrolytes – Na, K, CO_2, Cl, Ca, Mg, PO_4

Blood gases – pH, P_{CO_2}, P_{O_2}

Nonprotein nitrogenous substances – urea, uric acid, creatinine, ammonia

Bilirubin and urobilinogen

Hemoglobin, myoglobin

Thyroid hormones and binding proteins – T_4, T_3, RT_3

Therapeutic drugs – digoxin, quinidine, lidocaine, procainamide, Dilantin (diphenylhydantoin) lithium, tricyclic antidepressants

Toxicology – alcohols, salicylates, barbiturates, carbon monoxide, opium alkaloids, tranquilizers, qualitative screening techniques

Goals

Comprehension of the underlying principles of commonly used analytic methods, relative merits, automation feasibility, precision levels, common interferences, and approximate reference ranges

Learning Aids

Bibliography

Carbohydrates

Passey RB et al: Evaluation and comparison of 10 glucose methods and the reference method recommended in the proposed product class standard (1974). Clin Chem 23:131–139, 1977

Pennock CA, et al: A comparison of autoanalyzer methods for the estimation of glucose in blood. Clin Chim Acta 48:193–201, 1973

Radin N, et al: Glucose: An Annotated Bibliography. Atlanta: Center for Disease Control, October 1977

Lipids

Allain CC, et al: Enzymatic determinations of total serum cholesterol. Clin Chem 20:470–475, 1974

Bachorik PS, et al: Plasma high-density lipoprotein cholesterol concentrations determined after removal of other lipoproteins by heparin/manganese precipitation or by ultracentrifugation. Clin Chem 22:1828–1834, 1976

Bucolo G, David H: Quantitative determination of serum triglycerides by the use of enzymes. Clin Chem 19:476–482, 1973

Carter T, Wilding P: Factors involved in the determination of triglycerides in serum: An international study. Clin Chim Acta 70:433–447, 1976

Fasce CF, Vanderlinde RE: Factors affecting the results of serum cholesterol determinations: An interlaboratory evaluation. Clin Chem 18:901–908, 1972

Hatch FT, et al: Quantitative agarose gel electrophoresis of plasma lipoproteins: A simple technique and two methods for standardization. J Lab Clin Med 81:946–960, 1973

Zak B: Cholesterol methodologies: A review. Clin Chem 23: 1201–1214, 1977

Proteins

Doumas BT: Standards for total serum protein assays—a collaborative study. Clin Chem 21:1159–1166, 1975

Peters T: Serum albumin. Adv Clin Chem 13:37–111, 1970

Slater L, Carter PM, Hobbs JR: Measurement of albumin in the sera of patients. Ann Clin Biochem 12:33–40, 1975

Verbruggen R: Quantitative immunoelectrophoretic methods: A literature survey. Clin Chem 21:5–43, 1975

Enzymes

Amador E, Salvatore AC: Serum glutamic oxalacetic transaminase activity: Revised manual and automated methods using diazonium dyes. Am J Clin Pathol 55:686–697, 1971

Bowers GN, McComb RB: Measurement of total alkaline phosphatase activity in human serum. Clin Chem 21:1988–1995, 1975

Brydon WG, Smith AF: An appraisal of routine methods for the determination of the anodal isoenzymes of lactate dehydrogenase. Clin Chim Acta 43:361–369, 1973

Christensen HN, Palmer GA: Enzyme Kinetics: A Learning Program for Students of the Biological and Medical Sciences. Philadelphia: WB Saunders Co, 1974

Crowley LV, Alton M: A comparison of four methods for measuring creatine phosphokinase. Am J Clin Pathol 53:948–955, 1970

Morin LG: Evaluation of current methods for creatine kinase isoenzyme fractionation. Clin Chem 23(pt 1):205–210, 1977

Rodgerson DO, Osberg IM: Sources of error in spectrophotometric measurement of aspartate aminotransferase and alanine aminotransferase in serum. Clin Chem 20:43–50, 1974

Rosalki SB: Gamma-glutamyl transpeptidase. Adv Clin Chem 17: 53–107, 1975

Rosalki SB, Tarlow D: Amylase determination using insoluble substrates. Ann Clin Biochem 10:47–52, 1973

Searcy RL, Wilding P, Berk JF: An appraisal of methods for serum amylase determination. Clin Chim Acta 15:189–197, 1967

Tietz NW, Weinstock A, Rodgerson DO (ed): Proceedings of the Second International Symposium on Clinical Enzymology. Washington, DC: American Association of Clinical Chemistry, 1976

Wilkinson JH: Principles and Practices of Diagnostic Enzymology. Chicago: Year Book Medical Publishers Inc, 1976

Electrolytes

Schwartz HD, McConville BC, Christopherson EF: Serum ionized calcium by specific ion electrode. Clin Chim Acta 31:97–107, 1971

Blood Gases

Burnett RW, Noonan DC: Calculations and correction factors used in determination of blood pH and blood gases. Clin Chem 20: 1499–1506, 1974

Hill DW, Tilsley C: A comparative study of the performance of five commercial blood-gas and pH electrode analysers. Br J Anaesth 45: 647–654, 1973

Nonprotein Nitrogenous Substances

Cook JGH: Factors influencing the assay of creatinine. Ann Clin Biochem 12:219–232, 1975

Radin N, et al: Uric acid: An annotated bibliography. Atlanta: Center for Disease Control, 1974

Bilirubin

Hertz H: Bilirubins in human serum. Dan Med Bull 23:51–72, 1976

Hertz H, Dybkaer R, Lauritzen M: Direct spectrophotometric determination of the concentration of bilirubin in serum. Scand J Clin Lab Invest 33:215–230, 1974

Hemoglobin, Myoglobin

Rosano TG, Kenny MA: A radioimmunoassay for human serum myoglobin: Method development and normal values. Clin Chem 23: 69–75, 1977

Thyroid Hormones

McGowan GK: The laboratory assessment of thyroid function. Nomenclature. J Clin Pathol 28:207–254, 1975

Ratcliffe WA, Marshall J, Ratcliffe JG: The radioimmunoassay of 3, 3′, 5′-triiodothyronine (reverse T_3) in unextracted human serum. Clin Endocrinol 5:631–641, 1976

Wellby ML: The laboratory diagnosis of thyroid disorders. Adv Clin Chem 18:103–172, 1976

Therapeutic Drugs

Besch HR Jr, Watanabe AM: Radioimmunoassay of digoxin and digitoxin. Clin Chem 21:1815–1829, 1975

Marks V, Lindup WE, Baylis EM: Measurement of therapeutic agents in blood. Adv Clin Chem 16:47–109, 1973

Special issue on monitoring drugs in biological fluids. Clin Chem 22: 711–921, 1976

Toxicology

Special issue on toxicology and drug assay. Clin Chem 20:111–311, 1974

Texts (ASCP)

Council on Clinical Chemistry

Batsakis JG, Briere RO, Markel SF: Diagnostic Enzymology. Chicago: American Society of Clinical Pathologists, 1972 (Cat. no. 45-2-026-00—out of print)

Bittner DL, Gambino SR: Uric Acid Assays. Historical, Clinical and Current Concepts. Chicago: American Society of Clinical Pathologists, 1970 (Cat. no. 45-2-025-00—out of print)

Cooper GR, McDaniel V: Methods for a Determination of Glucose. Chicago: American Society of Clinical Pathologists, 1966 (Cat. no. 45-2-006-00—out of print)

Fleischer WR, Gambino SR: Blood pH, Po_2 and Oxygen Saturation. Chicago: American Society of Clinical Pathologists, 1972 (Cat. no. 45-2-029-00—out of print)

Gambino SR: Hemoglobinometry. Chicago: American Society of Clinical Pathologists, 1967 (Cat. no. 45-2-007-00—out of print)

Gambino SR, Di Re J: Bilirubin Assay. Chicago: American Society of Clinical Pathologists, 1968 (Cat. no. 45-2-023-00—out of print)

Passey RB: Problems in enzyme analysis, in Fuller JB (ed): Selected Topics in Clinical Chemistry II. Chicago: American Society of Clinical Pathologists, 1975, pp 25–42 (Cat. no. 45-2-033-00—out of print)

Other Councils

Penn GM, Davis T: Identification of Myeloma Proteins. Chicago: American Society of Clinical Pathologists, 1975 (Cat. no. 45-A-002-00)

Woodward SC, Hansell JR, Bering NM: The Determinations of T-3 and T-4. Chicago: American Society of Clinical Pathologists, 1973 (Cat. no. 45-8-005-00—out of print)

Audiovisual Aids (ASCP)

Finley PR: Digoxin Assay: Laboratory and Clinical Aspects. Chicago: American Society of Clinical Pathologists, 1978 (Cat. no. 47-8-027-00)

Fleischer WR, Hartmann AE: Blood Gases and Their Measurement—I. Chicago: American Society of Clinical Pathologists, 1976 (Cat. no. 47-2-022-00)

Fleischer WR, Hartmann AE: Blood Gases and Their Measurement—II. Chicago: American Society of Clinical Pathologists, 1976 (Cat. no. 47-2-024-00)

Fleischer WR, Weisberg H: Blood Gases and Pulmonary Dysfunction. Audiotape 38. Chicago: American Society of Clinical Pathologists, 1973 (Cat. no. 18-0-038-06—out of print)

Sleeper C: Laboratory and Clinical Use of Alkaline Phosphatase Isoenzymes. Audiotape 43. Chicago: American Society of Clinical Pathologists, 1973 (Cat. no. 18-0-043-06—out of print)

Tietz NW: Clinical Enzymology. Audiotape 43. Chicago: American Society of Clinical Pathologists, 1974 (Cat. no. 18-0-043-06—out of print)

Miscellaneous Learning Aids (ASCP)

Bayse D, Radin N, Lewis S: Calcium methodology evaluation and review, in Baer DM (ed): Technical Improvement Service. Number 17. Shilling JM (section ed). Chicago: American Society of Clinical Pathologists, 1974, pp 17–33 (Cat. no. 63-0-017-00—out of print)

Gambino SR: Phosphatase, alkaline; sodium; potassium; chloride; in Gambino SR (ed): Selected Clinical Chemistry "Check Sample" Critiques. Chicago: American Society of Clinical Pathologists, 1967 (Cat. no. 45-2-015-00—out of print)

Gambino SR (ed): Amylase; lipase; GOT; LDH; BUN and creatinine; glucose; in Selected Clinical Chemistry "Check Sample" Critiques, vol 2. Chicago: American Society of Clinical Pathologists, 1975 (Cat. no. 45-2-031-00—out of print)

Henry JB: Transaminase—SGOT, in Gambino SR (ed): Selected Clinical Chemistry "Check Sample" Critiques. Chicago: American Society of Clinical Pathologists, 1967, pp 167–173 (Cat. no. 45-2-015-00—out of print)

Straus R, Wurm M: Cholesterol, in Gambino SR (ed): Selected Clinical Chemistry "Check Sample" Critiques. Chicago: American Society of Clinical Pathologists, 1967 (Cat. no. 45-2-015-00—out of print)

Tietz NW: Clinical enzymology—present status and trends, in Baer DM (ed): Technical Improvement Service. Number 14. Chicago: American Society of Clinical Pathologists, 1973, pp 37–53 (Cat. no. 63-0-014-00—out of print)

Zettner A, Gambino SR: Calcium, in Gambino SR (ed): Selected Clinical Chemistry "Check Sample" Critiques. Chicago: American Society of Clinical Pathologists, 1967 (Cat. no. 45-2-015-00—out of print)

LEVEL II

Content

Carbohydrates: derivatives and metabolites—lactate, pyruvate, beta-hydroxybutyrate, urinary sugar identification

Lipids and lipoproteins—lipoprotein fractionation, phospholipid

Proteins—alpha$_1$-antitrypsin, complement, carcinoembryonic antigen (CEA), mucoproteins, paraprotein identification

Enzymes—aldolase, ceruloplasmin, 5′nucleotidase, arylamidase, cholinesterase, transketolase, trypsin, uropepsin, diamine oxidase

Polypeptide hormones

Steroid hormones

Catecholamines, serotonin and metabolites, prostaglandins

Porphyrins

Trace metals

Vitamins

Therapeutic drugs—anesthetic agents, antihypertensives, steroids, chemotherapeutic drugs, diuretics

Toxicology—specific identification of gases, volatiles, barbiturates, sedatives, hypnotics, heavy metals, cholinesterase inhibitors

Goals

Understanding of the basic principles, precision levels, common interferences, approximate reference ranges, automation feasibility, and the relative merits of uncommon or comparatively difficult analytic techniques

Learning Aids

Bibliography

Carbohydrates

Pennock CA: A review and selection of simple laboratory methods used for the study of glycosaminoglycan excretion and the diagnosis of the mucopolysaccharidoses. J Clin Pathol 29:111–123, 1976

Lipids and Lipoproteins

Burstein M, Scholnick HR, Morfin R: Rapid method for the isolation of lipoproteins from human serum by precipitation with polyanions. J Lipid Res 11:583–595, 1970

Wilson DE, Spiger MJ: A dual precipitation method for quantitative plasma lipoprotein measurement without ultracentrifugation. J Lab Clin Med 82:473–482, 1973

Proteins

Dietz AA, Rubinstein HM, Hodges L: Measurement of alpha$_1$-antitrypsin in serum, by immunodiffusion and by enzymatic assay. Clin Chem 20:396–399, 1974

Fleisher M, et al: Measurement of carcinoembryonic antigen. Clin Chem 19:1214–1220, 1973

Enzymes

Boyett JD, Lehmann HP, Beeler MF: Automated assay of ceruloplasmin by kinetic analysis of o-dianisidine oxidation. Clin Chim Acta 69:233–241, 1976

Garry PJ: Serum cholinesterase variants: Examination of several differential inhibitors, salts, and buffers used to measure enzyme activity. Clin Chem 17:183–191, 1971

Goldberg DM, Ellis G: Routine determination of 5′-nucleotidase activity of human serum using the LKB 8600 reaction rate analyzer. J Clin Pathol 25:907–909, 1972

Ismail AA, Williams DG: Scope and limitations of a kinetic assay for serum 5′nucleotidase activity. Clin Chim Acta 5:211–216, 1974

Polypeptide Hormones

Brown GM, Kirpalani SH: A critical view of the clinical relevance of growth hormone and its measurement in the nuclear medicine laboratory. Semin Nucl Med 5:273–285, 1975

Kleerekoper M, et al: Parathyroid hormone assay in primary hyperparathyroidism: Experiences with a radioimmunoassay based on commercially available reagents. Clin Chem 20:369–375, 1974

Malarkey WB: Recently discovered hypothalamic-pituitary hormone. Clin Chem 22:5–15, 1976

Porres JM, et al: Comparison of eight kits for the diagnosis of pregnancy. Am J Clin Pathol 64:452–463, 1975

Snider RH, et al: Immunochemical heterogeneity of calcitonin in man: Effect on radioimmunoassay. Clin Chim Acta 76:1–14, 1977

Steroid Hormones

Abraham GE: Radioimmunoassay of steroids in biological materials. Acta Endocrinol 183(suppl):1–42, 1974

Van de Calseyde JR, et al: Profiling urinary steroids. A reliable procedure. Clin Chim Acta 38:103–111, 1972

Wong P-Y, Wood DF, Johnson T: Routine radioimmunoassay of plasma testosterone, and results for various endocrine disorders. Clin Chem 21:206–210, 1975

Catecholamines

Goldenberg H: Specific photometric determination of 5-hydroxyindoleacetic acid in urine. Clin Chem 19:38–44, 1973

Knight JA, Fronk S, Haymond RE: Chemical basis and specificity of chemical screening tests for urinary vanilmandelic acid. Clin Chem 21:130–133, 1975

Porphyrins
Chisolm JJ, Brown DH: Micro-scale photofluorometric determination of 'free erythrocyte porphyrin' (protoporphyrin IX). Clin Chem 21: 1669–1682, 1975

Granick S, et al: Assays for porphyrins, delta-aminolevulinic-acid dehydratase, and porphyrinogen synthetase in microliter samples of whole blood: Applications to metabolic defects involving the heme pathway. Proc Natl Acad Sci USA 69:2381–2385, 1972

Uddin DE, Hicks JM (eds): Symposium on porphyrin measurements: Laboratory and clinical aspects. Clin Chem 23:251–274, 1977

Trace Metals
King JS (ed): Special issue on trace elements in clinical chemistry. Clin Chem 21:467–635, 1975

Vitamins
Demetriou JA: Vitamins, in Henry RJ, Cannon DC, Winkelman JW (eds): Clinical Chemistry: Principles and Technics. New York: Harper & Row Publishers Inc, 1974

Therapeutic Drugs
Davies DS: Biological Effects of Drugs in Relation to Their Plasma Concentrations. Baltimore: University Park Press, 1973

Gibaldi M, Perrier D: Pharmacokinetics. New York: Marcel Dekker Inc, 1975

Levy G (ed): Clinical Pharmacokinetics: A symposium. Washington, DC: American Pharmaceutical Association, Oct 1974

Notari RE: Biopharmaceutics and Pharmacokinetics: An Introduction, ed 2. New York: Marcel Dekker Inc, 1975

Shand DG: Drug therapy: Propranolol. N Engl J Med 293:280–285, 1975

Toxicology
Sunshine I: Methodology for Analytical Toxicology. Cleveland: CRC Press Inc, 1975

Texts (ASCP)

Council on Clinical Chemistry
Spikes JJ, Urry FM: Toxicology for the medical examiner, in Fuller JB (ed): Selected Topics in Clinical Chemistry. Chicago: American

Society of Clinical Pathologists, 1975, pp 1–24 (Cat. no. 45-2-033-00–out of print)

Stewart TC, Freeman JA: Vanilmandelic Acid and Catecholamine Determinations. Chicago: American Society of Clinical Pathologists, 1976 (Cat. no. 45-2-034-00)

Miscellaneous Learning Aids (ASCP)

Bell A: Hemoglobin S-C disease, in Baer DM (ed): Technical Improvement Service. Number 14. Chicago: American Society of Clinical Pathologists, 1973, pp 10–23 (Cat. no. 63-0-014-00–out of print)

Gambino SR (ed): Lithium; cortisol; in Selected Clinical Chemistry "Check Sample" Critiques, vol 2. Chicago: American Society of Clinical Pathologists, 1975 (Cat. no. 45-2-031-00–out of print)

LEVEL III

Content

Carbohydrates: derivatives and metabolites–automated high-resolution chromatographic identification of carbohydrate metabolites, organic anions, body fluids, and tissue preparations

Lipids and lipoproteins–tissue and cell homogenate biochemical techniques in atherosclerotic disease

Proteins and amino acids–high-resolution automated chromatographic identification of protein metabolites and amino acids in body fluids, tissues and cell homogenates; techniques for the identification of inborn errors of protein metabolism

Abnormal hemoglobins–techniques for specific characterization

Enzymes–tissue techniques for inborn errors of metabolism

Broad spectrum toxicological analysis; pediatric toxicology–soaps, detergents, insecticides, paints, polishes, waxes

Goals

Expertise in one or more areas of chemical analysis, and proficiency in the highly specialized analytic methods

Learning Aids

Bibliography

Constantopoulos G, Dekaban AS: Chemical definition of the mucopolysaccharidoses. Clin Chim Acta 59:321–336, 1975

Hosli P: Quantitative assays of enzyme activity in single cells: Early prenatal diagnosis of genetic disorders. Clin Chem 23:1476–1484, 1977

Kissinger PT, et al: Recent developments in the clinical assessment of the metabolism of aromatics by high-performance, reversed-phase chromatography with amperometric detection. Clin Chem 23: 1449–1455, 1977

Norum KR, Gjone E: Lecithin: cholesterol acyltransferase: Recent research on biochemistry and physiology of the enzyme. Scand J Clin Lab Invest 33:191–197, 1974

Seidel D, Alaupovic P, Furman RH: A lipoprotein characterizing obstructive jaundice. I. Method for quantitative separation and identification of lipoproteins in jaundiced subjects. J Clin Invest 48: 1211–1223, 1969

FUNCTIONAL TESTS

LEVEL I

Content

Carbohydrate tolerance

Hepatic function

Renal function

Thyroid function

Adrenocortical function

Intestinal absorption and pancreatic exocrine functions

Goals

Ability to formulate a reasonable working protocol for each test through an understanding of the underlying principles and of pathophysiology

Learning Aids

Bibliography

Carbohydrate Tolerance

Klimt CR, et al: Standardization of the glucose tolerance test: Report of the Committee on Statistics of the American Diabetes Association. Diabetes 18:299–305, 1968

Seltzer AS: Oral glucose tolerance tests, in Fajans SS, Sussman KE (eds): Diabetes Mellitus: Diagnosis and Treatment. New York: American Diabetes Association Inc, 1971

Hepatic Function

Burke MD: Liver function. Hum Pathol 6:273–286, 1975

Renal Function

Kassirer JP: Clinical evaluation of kidney function—glomerular function. N Engl J Med 285:385–389, 1971

Kassirer JP: Clinical evaluation of kidney function—tubular function. N Engl J Med 285:499–502, 1971

Mitchell FL, Veall N, Watts RWE: Renal function tests suitable for clinical practice. Ann Clin Biochem 9:3–20, 1972

Morgan DB, Dillon S, Payne RB: The assessment of glomerular function: Creatinine clearance or plasma creatinine? Postgrad Med J 54:302–310, 1978

Thyroid Function

Alsever RN, Gotlin RW: The thyroid, in Handbook of Endocrine Tests, ed 2. Chicago: Year Book Medical Publishers Inc, 1978, pp 43–60

Adrenocortical Function

Eddy RL, et al: Cushing's syndrome, a prospective study of diagnostic methods. Am J Med 55:621–630, 1973

Nichols T, Nugent CA, Tyler FH: Steroid laboratory tests in the diagnosis of Cushing's syndrome. Am J Med 45:116–128, 1968

Tyler RH, West CD: Laboratory evaluation of disorders of the adrenal cortex. Am J Med 53:664–672, 1972

Intestinal Absorption and Pancreatic Exocrine Functions
Gowenlock AH: Tests of exocrine pancreatic function. Ann Clin Biochem 14:61–89, 1977

Russell RI, Lee FD: Tests of small intestinal function—digestion, absorption, secretion. Clin Gastroenterol 7:277–315, 1978

Wormsley KG: Tests of pancreatic secretion. Clin Gastroenterol 7: 529–544, 1978

Texts (ASCP)

Council on Clinical Chemistry
Bittner DL, Gambino SR: Thyroid Disease: PBI, T4, Free Thyroxine or T3. Chicago: American Society of Clinical Pathologists, 1968 (Cat. no. 45-2-024-00—out of print)

Clerch AR: Adrenal Cortex Function. Chicago: American Society of Clinical Pathologists, 1972 (Cat. no. 45-2-027-00—out of print)

Cooper GR, McDaniel V: Methods for the Determination of Glucose. Chicago: American Society of Clinical Pathologists, 1970 (Cat. no. 45-2-006-00—out of print)

Fuller JB (ed): Liver Function Tests. Chicago: American Society of Clinical Pathologists, 1966 (Cat. no. 45-2-010-00—out of print)

Other Councils
Woodward SC, Hansell JR, Bering NM: The Determinations of T-3 and T-4. Chicago: American Society of Clinical Pathologists, 1973 (Cat. no. 45-8-005-00—out of print)

LEVEL II

Content

Fetoplacental function

Hypothalamic-pituitary function

Gonadal function

Gastrointestinal and pancreatic endocrine functions

Goals

Ability to formulate a working protocol for more complicated, less commonly used functional tests, and to assess the relative merits of different approaches

Learning Aids

Bibliography

Fetoplacental Function

Doran TA, et al: Amniotic fluid tests for fetal maturity. Am J Obstet Gynecol 119:829–837, 1974

Wilde CE, Oakey RE: Biochemical tests for the assessment of fetoplacental function. Ann Clin Biochem 12:83–118, 1975

Hypothalamic-pituitary Function

Alsever RN, Gotlin RW: Handbook of Endocrine Tests in Adults and Children. Chicago: Year Book Medical Publishers Inc, 1975

Archer DF: Current concepts of prolactin physiology in normal and abnormal conditions. Fertil Steril 28:125–134, 1977

Eddy RL, et al: Human growth hormone release. Am J Med 56: 179–185, 1974

Lin TU, Tucci JR: Provocative tests of growth-hormone release: A comparison of results with seven stimuli. Ann Intern Med 80: 464–469, 1974

Gonadal Function

Alsever RN, Gotlin RW: Handbook of Endocrine Tests in Adults and Children. Chicago: Year Book Medical Publishers Inc, 1975

Gastrointestinal and Pancreatic Endocrine Functions

Horwitz DL, Kuzuya H, Rubenstein AH: Circulating serum C-peptide: A brief review of diagnostic implications. N Engl J Med 295: 207–209, 1976

Piper DW: Tests of gastric function. Clin Gastroenterol 7:247–276, 1978

Rayford PL, Miller TA, Thompson JC: Secretin, cholecystokinin and newer gastrointestinal hormones. N Engl J Med 294:1093–1101, 1157–1163, 1976

Schein PS, DeLellis RA, Kahn CR: Islet-cell tumors: Current concepts and management. Ann Intern Med 79:239–257, 1973

Straus E: Radioimmunoassay of gastrointestinal hormones. Gastroenterol 74:141–152, 1978

Walsh JH, Grossman MI: Gastrin. N Engl J Med 292:1324–1334, 1377–1384, 1975

Text (ASCP)

Dito WR, Patrick CW, Shelly J: Clinical Pathologic Correlations in Amniotic Fluid. Chicago: American Society of Clinical Pathologists, 1975 (Cat. no. 45-9-005-00)

Audiovisual Aid (ASCP)

Dito WR: Amniotic Fluid Analysis in Pregnancy at Risk. Chicago: American Society of Clinical Pathologists, 1975 (Cat. no. 47-9-001-00)

Miscellaneous Learning Aid (ASCP)

Gambino SR (ed): Estriol, in Selected Clinical Chemistry "Check Sample" Critiques, vol. 2. Chicago: American Society of Clinical Pathologists, 1975 (Cat. no. 45-2-031-00—out of print)

SPECIAL ANALYTIC METHODS

LEVEL I

Content

Competitive-binding assay—competitive protein-binding assay and radioimmunoassay

Immunochemical techniques—immunodiffusion, immunoelectrophoresis, nephelometry

Goals

Sufficient understanding of commonly used immunochemical techniques and competitive-binding assays to be able to evaluate the relative merits of different commercial kits and reagents, and to initiate or modify a method

Learning Aids

Bibliography

Competitive-binding Assays

Cameron EHD, Hillier SG, Griffiths K (eds): Steroid Immunoassay: Proceedings of the Fifth Tenovus Workshop. Cardiff, Wales: Alpha Omega, 1975

Freeman LM, Blaufox MD (eds): Radioimmunoassay. New York: Grune & Stratton Inc, 1975

Odell WD, Daughaday WH (eds): Principles of Competitive Protein-binding Assays. Philadelphia: JB Lippincott Co, 1971

Immunochemical Techniques

Fahey JL, McKelvey EM: Quantitative determination of serum immunoglobulins in antibody-agar plates. J Immunol 94:84–90, 1965

Mancini G, Carbonara AP, Heremans JF: Immunochemical quantitation of antigens by single radial immunodiffusion. Int J Immunochem 2:235–254, 1965

Ouchterlony O: Handbook of Immunodiffusion and Immunoelectrophoresis. Ann Arbor, Mich: Ann Arbor Science Publishers Inc, 1968

Virella G, Fudenberg HH: Comparison of immunoglobulin determinations in pathologic sera by radial immunodiffusion and laser nephelometry. Clin Chem 23:1925–1928, 1977

Texts and Atlases (ASCP)

Council on Clinical Chemistry

Cawley LP: Electrophoresis and Immunoelectrophoresis. Chicago: American Society of Clinical Pathologists, 1966 (Cat. no. 45-2-009-00–out of print)

Cawley LP, et al: Basic Electrophoresis, Immunoelectrophoresis and Immunochemistry. Chicago: American Society of Clinical Pathologists, 1972 (Cat. no. 45-2-030-00–out of print)

Cawley LP, Minard BJ, Penn GM: Electrophoresis and Immunochemical Reactions in Gels: Techniques and Interpretation. Chicago: American Society of Clinical Pathologists, 1978 (Cat. no. 45-2-035-00)

Other Councils

Cawley LP, et al: Clinical Immunology and Immunochemistry. Chicago: American Society of Clinical Pathologists, 1972 (Cat. no. 45-A-001-00—out of print)

Penn GM, Batya J: Interpretation of Immunoelectrophoretic Patterns. Chicago: American Society of Clinical Pathologists, 1975 (Atlas and slides: Cat. no. 15-A-001-00/Atlas only: Cat. no. 16-A-001-00)

Penn GM, Davis T: Identification of Myeloma Proteins. Chicago: American Society of Clinical Pathologists, 1975, (Cat. no. 45-A-002-00)

Reynoso G: Competitive Protein Binding and Radioimmunoassay. Chicago: American Society of Clinical Pathologists, 1972 (Cat. no. 45-8-004-00—out of print)

Audiovisual Aids (ASCP)

Cawley LP: Principles of Radioimmunoassay—Part I. Chicago: American Society of Clinical Pathologists, 1975 (Cat. no. 47-8-009-00)

Cawley LP: Principles of Radioimmunoassay—Part II. Chicago: American Society of Clinical Pathologists. 1975 (Cat. no. 47-8-010-00)

Glassman A: Quantitative Immunodiffusion. Audiotape 33. Chicago: American Society of Clinical Pathologists, 1971 (Cat. no. 18-0-033-06—out of print)

Reynoso G, Cawley LP: The Application and Potential of Radioimmunoassay. Audiotape 38. Chicago: American Society of Clinical Pathologists, 1973 (Cat. no. 18-0-038-06—out of print)

LEVEL II

Content

Competitive-binding assay—radioreceptor assay, enzyme-linked immunosorbent assay (ELISA), enzyme-multiplied immunoassay (EMIT), free radical assay (FRAT)

Noncompetitive-binding assay—immunoradiometric assay

Goals

Ability to initiate or modify competitive-binding and noncompetitive-binding assays that may have more practical application in the future, and to compare and evaluate the commercial kits and reagents associated with these assays

Learning Aids

Bibliography

Competitive-binding Assays

Kahn CR, Roth J: Cell membrane receptors for polypeptide hormones. Am J Clin Pathol 63:656–667, 1975

Voller A, Bartlett A, Bidwell DE: Enzyme immunoassays with special reference to ELISA techniques. J Clin Pathol 31:507–520, 1978

Walker WH: An approach to immunoassay. Clin Chem 23:384–402, 1977

Wisdom GB: Enzyme-immunoassay. Clin Chem 22:1243–1255, 1976

Yolken RH: ELISA linked immunosorbent assay. Hospital Practice 13:121–127, 1978

Noncompetitive-binding Assays

Rodbard D, Weiss GH: Mathematical theory of immunoradiometric (labeled antibody) assays. Anal Biochem 52:10–44, 1973

ANALYSIS OF SPECIAL TYPES OF SPECIMENS

LEVEL I

Content

Spinal fluid

Amniotic fluid

Joint fluid

Gastric fluid

Intestinal fluid

Feces

Calculi

Goals

Ability to analyze specimens (other than blood and urine) that are subjected to regular biochemical measurements through precise knowledge of patient preparation and sampling methods, familiarity with the principles and limitations of commonly used techniques, and a knowledge of the principles of polarizing microscopy for the analysis of joint fluid

Learning Aids

Bibliography

Spinal Fluid

Johnson KP, Nelson BJ: Multiple sclerosis: Diagnostic usefulness of cerebrospinal fluid. Annals of Neurology 2:425–431, 1977

Lamoureux G, et al: Cerebrospinal fluid proteins in multiple sclerosis. Neurology 25:537–546, 1975

Patten BM: How much blood makes the CSF bloody? JAMA 206: 378, 1968

Savory J, Brody JP: Measurement and diagnostic value of cerebrospinal fluid enzymes. Ann Clin Lab Sci 9:68–79, 1979

Amniotic Fluid

Burnett RW: Instrumental and procedural sources of error in determination of bile pigments in amniotic fluid. Clin Chem 18:150–154, 1972

Kapitulnik J, Kaufmann NA, Blondheim SH: Chemical versus spectrophotometric determination of bilirubin in amniotic fluid and the influence of hemoglobin and methene pigments. Clin Chem 16: 756–759, 1970

Natelson S, Scommegna A, Epstein MB: Amniotic Fluid: Physiology, Biochemistry, and Clinical Chemistry. New York: John Wiley & Sons Inc, 1974

Joint Fluid

Phelps P, Steele AD, McCarty DJ: Compensated polarized light microscopy: Identification of crystals in synovial fluid from gout and pseudogout. JAMA 203:508–512, 1968

Gastric Fluid

Piper DW: Tests of gastric function. Clin Gastroenterol 7:247–276, 1978

Intestinal Fluid

Russell RI, Lee FD: Tests of small intestinal function: digestion, absorption, secretion. Clin Gastroenterol 7:277–315, 1978

Feces

Morris DW, et al: Reliability of chemical tests for fecal occult blood in hospitalized patients. Am J Dig Dis 21:845–852, 1976

Calculi

Fiereck EA: Analysis of calculi, in Tietz NW (ed): Fundamentals of Clinical Chemistry, ed 2. Philadelphia: WB Saunders Co, 1976, pp 1015–1025

Kleeberg J: Simplified qualitative chemical analysis for urinary calculi. J Clin Pathol 29:1038–1039, 1976

Text (ASCP)

Dito WR, Patrick CW, Shelly J: Clinical Pathologic Correlations in Amniotic Fluid. Chicago: American Society of Clinical Pathologists, 1975 (Cat. no. 45-9-005-00)

Audiovisual Aids (ASCP)

Dito WR: Amniotic Fluid Analysis in Pregnancy at Risk. Chicago: American Society of Clinical Pathologists, 1975 (Cat. no. 47-9-001-00)

Miscellaneous Learning Aid (ASCP)

Gambino SR (ed): Renal calculi, in Selected Clinical Chemistry "Check Sample" Critiques, vol 2. Chicago: American Society of Clinical Pathologists, 1975 (Cat. no. 45-2-031-00)—out of print)

LEVEL II

Content

Amniotic fluid

Calculi

Tissue and cell homogenates

Goals

Precise, working knowledge of the handling of amniotic fluid and of cellular preparations for the detection of inborn errors of metabolism; and an understanding of the optical, infrared, and x-ray diffraction methods involved in stone analysis

Learning Aids

Bibliography

Amniotic Fluid

Nadler HL: Prenatal diagnosis of inborn defects: A status report. Hospital Practice 10:41–51, 1975

Rennert OM: Antenatal diagnosis of genetic disease. Ann Clin Lab Sci 5:153–160, 1975

Screening for neural-tube defects, editorial. Lancet 1:1345–1346, 1977

Calculi

Morriss RH, Beeler MF: X-ray diffraction analysis of 464 urinary calculi. Am J Clin Pathol 48:413–417, 1967

Tsay YC: Applications of infrared spectroscopy to analysis of urinary calculi. J Urol 86:838–854, 1961

Tissue and Cell Homogenates

Rennert OM: Syndrome of the defective lysosome—the genetic mucopolysaccharides. Ann Clin Lab Sci 5:355–362, 1975

Thomas GH, Scott CI: Laboratory diagnosis of genetic disorders. Pediatr Clin North Am 20:105–119, 1973

Section 2

CHEMICAL PATHOLOGY

DISORDERS OF CARBOHYDRATE METABOLISM

LEVEL I

Content

Pathophysiology—disordered glucose homeostasis

Diagnosis—diabetes mellitus and metabolic decompensation in the diabetic patient; diagnostic strategies for suspected hypoglycemia in the adult patient

Goals

Sufficient knowledge of the pathophysiology of glucose homeostasis and of the clinical aspects of diabetes mellitus and hypoglycemia to detect and diagnose specifically diabetes mellitus and to outline a strategy for the evaluation of suspected hypoglycemia

Learning Aids

Bibliography

Factitious hypoglycemia, editorial. Lancet 1:1293, 1978

Fajans SS, Floyd JC: Fasting hypoglycemia in adults. N Engl J Med 294:766–772, 1976

Irvine WJ: Classification of idiopathic diabetes. Lancet 1:638–642, 1977

Jarrett RJ, Keen H: Hyperglycemia and diabetes mellitus. Lancet 2: 1009–1012, 1976

Marks V: Hypoglycemia. 2. Other causes. Clin Endocrinol Metabol 5: 769–782, 1976

Marks V, Alberti KG: Suggested tests of carbohydrate metabolism. Clin Endocrinol Metabol 5:805–820, 1976

Merimee TJ, Tyson JE: Stabilization of plasma glucose during fasting. N Engl J Med 291:1275–1278, 1974

Newsholme EA: Carbohydrate metabolism in vivo: Regulation of the blood glucose level. Clin Endocrinol Metabol 5:543–578, 1976

Nuttall FQ: Diagnosis of mild diabetes. Minn Med 59:747–752, 1976

Prout TE: The use of screening and diagnostic procedures: The oral glucose tolerance test. Md State Med J 25:62–65, 1976

Reynolds C, Orchard DB: The oral glucose tolerance test revisited and revised. Can Med Assoc J 116:1223–1224, 1977

Rifkin H: Why control diabetes? Med Clin North Am 62:747–752, 1978

Siperstein MD: The glucose tolerance test: A pitfall in the diagnosis of diabetes mellitus. Adv Intern Med 20:297–323, 1975

Sönksen PH, Judd SL, Lowy C: Home monitoring of blood glucose: Method for improving diabetic control. Lancet 1:729–732, 1978

Turner RC, Holman RR: Insulin rather than glucose homeostasis in the pathophysiology of diabetes. Lancet 1:1272–1274, 1976

Unger RH: Alpha- and beta-cell interrelationships in health and disease. Metabolism 23:581–593, 1974

Unger RH, Orci L: The essential role of glucagon in the pathogenesis of diabetes mellitus. Lancet 1:14–16, 1975

Whitehouse FW: The diagnosis of diabetes: How to determine which patients to treat. Med Clin North Am 62:627–637, 1978

Williams RH: Etiologic, pathophysiologic, and clinical interrelationships in diabetes. Johns Hopkins Med J 136:25–37, 1975

Text (ASCP)

Council on Clinical Chemistry

Cooper GR, McDaniel V: Methods for the Determination of Glucose. Chicago: American Society of Clinical Pathologists, 1970 (Cat. no. 45-2-006-00—out of print)

LEVEL II

Content

Pathophysiology—disordered carbohydrate metabolism

Diagnosis—test strategies for hypoglycemic disorders in infants and children; management strategies

Goals

Sufficient knowledge of errors of carbohydrate metabolism to devise test strategies for hypoglycemic disorders in neonates and children, and adequate understanding to give advice on interpretation of test results and patient management

Learning Aids

Bibliography

Behrman RE: Neonatology: Diseases of the Fetus and Infant. St Louis: CV Mosby Co, 1973, pp 439–443

Cahill GF, Soeldner JS: A non-editorial on non-hypoglycemia. N Engl J Med 291:905–906, 1974

Diagnosis of insulinoma, editorial. Lancet 2:385–386, 1974

Dweck HS: Neonatal hypoglycemia and hyperglycemia. Postgrad Med 60:118–124, 1976

Factitious hypoglycemia, editorial. Lancet 1:1293, 1978

Frerichs H, Creutzfeldt W: Hypoglycemia: I. Insulin secreting tumors. Clin Endocrinol Metabol 5:747–767, 1976

Gonen B, Rubenstein AH: Hemoglobin AI and diabetes mellitus. Diabetologia 15:1–8, 1978

Horwitz DL, Kuzuya H, Rubenstein AH: Circulating serum C-peptide. N Engl J Med 295:207–209, 1976

Horwitz DL, Rubenstein AH, Steiner DF: Proinsulin and C-peptide diabetes. Med Clin North Am 62:723–733, 1978

Kalhan SC, Savin SM, Adam PAJ: Attenuated glucose production rate in newborn infants of insulin-dependent diabetic mothers. N Engl J Med 296:375–376, 1977

Koenig RJ, et al: Correlation of glucose regulation and hemoglobin AI_c. N Engl J Med 295:417–420, 1976

Kreisberg RA, Pennington LF: Tumor hypoglycemia: A heterogeneous disorder. Metabolism 19:445–452, 1970

Levine R: Mechanisms of insulin secretion. N Engl J Med 283: 522–526, 1970

Marks V: Alcohol and carbohydrate metabolism. Clin Endocrinol Metabol 7:333–349, 1978

Neonatal hypoglycemia and nesidioblastosis, editorial. Lancet 1: 193–194, 1978

Senior B: Neonatal hypoglycemia. N Engl J Med 289:790–793, 1973

Siperstein MD, et al: Control of blood glucose in diabetic vascular disease. N Engl J Med 296:1060–1063, 1977

LEVEL III

Content

Pathophysiology—new developments in the pathophysiology of carbohydrate disorders

Diagnosis—role of cellular biochemistry in the specific diagnosis of inborn errors of carbohydrate metabolism

Management—critical evaluation of the use of laboratory tests in the detection and management of diabetes mellitus

Goals

Advanced knowledge of cellular errors of carbohydrate metabolism; appreciation of new, scientific developments; ability to evaluate current and new approaches for the detection and management of diabetes mellitus

Learning Aids

Bibliography

Davidson MB: The case for control in diabetes mellitus. West J Med 129:193–206, 1978

Glucoreceptors, insulin release, and diabetes, editorial. Lancet 2: 646–647, 1975

Havel RJ: Caloric homeostasis and disorders of fuel transport. N Engl J Med 287:1186–1192, 1972

Kreisberg RA: Diabetic ketoacidosis: New concepts and trends in pathogenesis and treatment. Ann Intern Med 88:681–695, 1978

Levine R: Glucagon and the regulation of blood sugar. N Engl J Med 294:494–495, 1976

Melmed RN: Intermediate cells of the pancreas. An appraisal. Progress in Gastroenterology 76:196–201, 1978

Neufeld EF: The biochemical basis for mucopolysaccharidoses and mucolipidoses. Prog Med Genet 19:81–101, 1974

Steiner DF: Errors in insulin biosynthesis. N Engl J Med 294: 952–953, 1976

West KM: Substantial differences in the diagnostic criteria used by diabetes experts. Diabetes 24:641–644, 1975

Williamson JR, Kilo C: Current status of capillary basement-membrane disease in diabetes mellitus. Diabetes 26:65–73, 1977

Winegrad AI, Greene DA: Diabetic polyneuropathy: The importance of insulin deficiency, hyperglycemia and alterations in myoinositol metabolism in its pathogenesis. N Engl J Med 295:1416–1421, 1976

Young DS: High pressure column chromatography of carbohydrates in the clinical laboratory. Am J Clin Pathol 53:803–810, 1970

DISORDERS OF LIPID METABOLISM

LEVEL I

Content

Pathophysiology—disordered lipoprotein metabolism

Diagnosis—diagnostic strategy for the assessment of cardiovascular lipid risk

Goals

Sufficient knowledge of lipoprotein metabolism and cardiovascular risk to devise a diagnostic strategy using cholesterol, triglyceride, and plasma inspection

Learning Aids

Bibliography

Castelli WP: CHD risk factors in the elderly. Hospital Practice 11: 113–121, 1976

Corday E: Can we reverse or retard an obstructive coronary lesion by risk factor intervention? Cleve Clin Q 45:5–8, 1978

Fisher WR, Truitt DH: The common hyperlipoproteinemias: An understanding of disease mechanisms and their control. Ann Intern Med 85:497–508, 1976

Fredrickson DS: It's time to be practical. Circulation 51:209–211, 1975

Havel RJ: Classification of the hyperlipidemias. Annu Rev Med 28: 195–209, 1977

Iammarino RM: Lipoprotein electrophoresis should be discontinued as a routine procedure. Clin Chem 21:300–308, 1975

Kannel WB, Sorlie P: Utility of conventional risk factors in evaluation of patients with coronary disease. Cleve Clin Q 45:1–4, 1978

Levy RI: The meaning of lipid profiles. Postgrad Med 57:34–38, 1975

Lewis B: The common hyperlipidaemias. Postgrad Med J 52: 433–437, 1976

LEVEL II

Content

Pathophysiology—genetic hyperlipidemias

Diagnosis—special test strategies for hyperlipoproteinemic states

Goals

The ability to devise special test strategies for hyperlipoproteinemia that include low-density lipoprotein (LDL) and high-density lipoprotein (HDL) measurements

Learning Aids

Bibliography

Albers JJ, Cheung MC, Hazzard WR: High-density lipoproteins in myocardial infarction survivors. Metabolism 27:479–485, 1978

Brown MS, Goldstein JL: Familial hypercholesterolemia: A genetic defect in the low-density lipoprotein receptor. N Engl J Med 294: 1386–1390, 1976

Castelli WP, et al: HDL cholesterol and other lipids in coronary heart disease. The cooperative lipoprotein phenotyping study. Circulation 55:767–772, 1977

Fisher WR, Truitt DH: The common hyperlipoproteinemias: An understanding of disease mechanisms and their control. Ann Intern Med 85:497–508, 1976

Goldstein JL, Brown MS: Familial hypercholesterolemia: A genetic regulatory defect in cholesterol metabolism. Am J Med 58: 147–150, 1975

Goldstein JL, et al: Hyperlipidemia in coronary heart disease. II. Genetic analysis of lipid levels in 176 families, and delineation of a new inherited disorder, combined hyperlipidemia. J Clin Invest 52:1544-1568, 1973

Kannel WB, Castelli WP, Gordon T: Cholesterol in the prediction of atherosclerotic disease: New perspectives based on the Framingham study. Ann Intern Med 90:85-91, 1979

Leonard JV, et al: Screening for familial hyper-beta-lipoproteinemia in children in hospital. Arch Dis Child 51:842-847, 1976

Miller GJ: High-density lipoprotein, low-density lipoprotein, and coronary heart disease. Thorax 33:137-139, 1978

Motulsky AG: The genetic hyperlipidemias. N Engl J Med 294: 823-827, 1976

Tall AR, Small DM: Plasma high-density lipoproteins. N Engl J Med 299:1232-1236, 1978

LEVEL III

Content

Pathophysiology—the role of lipids in atherosclerosis; inborn errors of lipid metabolism

Diagnosis—critical analysis of the role of laboratory tests as risk-factor indicators for atheromatous disease; cellular biochemical test strategies for specific diagnosis in inborn errors of lipid metabolism

Goals

Advanced knowledge of inherited lipid disorders and of the role of lipids in atherosclerosis. Ability to devise cellular biochemical test strategies for specific diagnosis in inborn errors of lipid metabolism

Learning Aids

Bibliography

Albers JJ, Warnick GR, Hazzard WR: Type III hyperlipoproteinemia: A comparative study of current diagnostic techniques. Clin Chim Acta 75:193-204, 1977

Atherogenic hormones, editorial. Lancet 1:1347–1348, 1977

Bilheimer DW: Needed: New therapy for hypercholesterolemia. N Engl J Med 296:508–510, 1977

Brown MS, Goldstein JL: Familial hypercholesterolemia: A genetic defect in the low-density lipoprotein receptor. N Engl J Med 294: 1386–1390, 1976

The future in lipid diseases, editorial. Lancet 2:808, 1977

Ross R, Glomset JA: The pathogenesis of atherosclerosis. N Engl J Med 295:369–377, 420–425, 1976

Sodhi HS, Mason DT: New insights into the homeostasis of plasma cholesterol. A time for changing concepts. Am J Med 63:325–327, 1977

SERUM PROTEIN ABNORMALITIES

LEVEL I

Content

Pathophysiology—altered plasma protein homeostasis

Diagnosis—interpretation of altered serum total protein and albumin concentrations; indications for—and interpretation of—serum electrophoretic patterns

Goals

Understanding of altered plasma protein homeostasis; and the ability to interpret routine serum protein test results and serum electrophoretic patterns

Learning Aids

Bibliography

Alper CA: Plasma protein measurements as a diagnostic aid. N Engl J Med 291:287–290, 1974

Larson PH: Serum proteins: Diagnostic significance of electrophoretic patterns. Hum Pathol 5:629–640, 1974

Laurell CB: Electrophoresis, specific protein assays, or both in measurement of plasma protein? Clin Chem 19:99–102, 1973

Rothschild MA, Oratz M, Schreiber SS: Albumin metabolism. Gastroenterology 64:324–337, 1973

Rothschild MA, Oratz M, Schreiber SS: Albumin synthesis. N Engl J Med 286:748–757, 1972

Rothschild MA, Oratz M, Schreiber SS: Albumin synthesis (second of two parts). N Engl J Med 286:816–821, 1972

Sunderman FW: Studies on the serum proteins: VI. Recent advances in clinical interpretation of electrophoretic fractionations. Am J Clin Pathol 42:1–21, 1964

Texts (ASCP)

Cawley LP, et al: Basic Electrophoresis, Immunoelectrophoresis, and Immunochemistry. Chicago: American Society of Clinical Pathologists, 1972 (Cat. no. 45-2-030-00—out of print)

Cawley LP, Minard BJ, Penn GM: Electrophoresis and Immunochemical Reactions in Gels: Techniques and Interpretation. Chicago: American Society of Clinical Pathologists, 1978 (Cat. no. 45-2-035-00)

LEVEL II

Content

Pathophysiology—acute phase reactants; $alpha_1$-antitrypsin; fetal proteins

Diagnosis—$Alpha_1$-antitrypsin deficiency; the role of fetal proteins in diagnosis and management of disease

Goals

Understanding of the acute phase reaction, the detection and specific diagnosis of $alpha_1$-antitrypsin deficiency, the role of fetal proteins in diagnosis and management, and multivariate analysis of electrophoretic pattern test results

Learning Aids

Bibliography

Alpert E: Alpha$_1$-fetoprotein: Need for quantitative assays. N Engl J Med 290:568–569, 1974

Antitrypsin deficiency and adult cirrhosis, editorial. Lancet 1:925, 1973

Barkin JS, et al: Initial levels of CEA and their rate of change in pancreatic carcinoma following surgery, chemotherapy, and radiation therapy. Cancer 42 (suppl 3): 1472–1476, 1978

Galen RS, Gambino SR: Carcinoembryonic antigen (CEA) assay. Hum Pathol 6:128–130, 1975

Hansen HJ, et al: Carcinoembryonic antigen (CEA) assay. A laboratory adjunct in the diagnosis and management of cancer. Hum Pathol 5:139–147, 1974

Johnson PJ, Portmann B, Williams R: Alpha-fetoprotein concentrations measured by radioimmunoassay in diagnosing and excluding hepatocellular carcinoma. Br Med J 2:661–663, 1978

Kalser MH, et al: Circulating carcinoembryonic antigen in pancreatic carcinoma. Cancer 42 (suppl 3):1468–1471, 1978

Lieberman J: Alpha$_1$-antitrypsin deficiency. Med Clin North Am 57: 691–706, 1973

Moertel CG, Schutt AJ, Go VLW: Carcinoembryonic antigen test for recurrent colorectal carcinoma. JAMA 239:1065–1066, 1978

Murray-Lyon IM, et al: Prognostic value of serum alpha-fetoprotein in fulminant hepatic failure including patients treated by charcoal haemoperfusion. Gut 17:576–580, 1976

Nakamura RM, Plow EF, Edgington TS: Current status of carcinoembryonic antigen (CEA) and CEA-S assays in the evaluation of neoplasm of the gastrointestinal tract. Ann Clin Lab Sci 8:4–10, 1978

Rynbrandt DJ, Ihrig J, Kleinerman J: Serum trypsin inhibitory capacity and Pi phenotypes. I. Methods and control values. Am J Clin Pathol 63:251–260, 1975

Schultz H, et al: Serum alpha-fetoprotein and human chorionic gonadotropin as markers for the effect of postoperative radiation therapy and/or chemotherapy in testicular cancer. Cancer 42:2182–2186, 1978

Shani A, et al: Serial plasma carcinoembryonic antigen measurements in the management of metastatic colorectal carcinoma. Ann Intern Med 88:627–630, 1978

Sharp HL: Alpha-1-antitrypsin deficiency. Hospital Practice 6:83–96, 1971

Sugarbaker PH, Zamcheck N, Moore FD: Assessment of serial carcinoembryonic antigen (CEA) assays in postoperative detection of recurrent colorectal cancer. Cancer 38:2310–2315, 1976

Ubiquitous human chorionic gonadotrophin, editorial. Lancet 2:1116, 1977

Wanebo HJ, et al: Preoperative carcinoembryonic antigen level as a prognostic indicator in colorectal cancer. N Engl J Med 299: 448–451, 1978

Werner M, Brooks SH, Cohnen G: Diagnostic effectiveness of electrophoresis and specific protein assays, evaluated by discriminant analysis. Clin Chem 18:116–123, 1972.

Zeltzer PM, et al: Differentiation between neonatal hepatitis and biliary atresia by measuring serum-alpha-fetoprotein. Lancet 1: 373–375, 1974

Text (ASCP)

Council on Clinical Chemistry

Cawley LP: Electrophoresis and Immunoelectrophoresis. Chicago: American Society of Clinical Pathologists, 1966 (Cat. no. 45-2-009-00–out of print)

LEVEL III

Content

Pathophysiology–genetic polymorphism of serum proteins; disorders of amino acid metabolism

Goals

In-depth knowledge of amino acid metabolism and of the use of tissue and cellular homogenate biochemistry in specific diagnosis

Learning Aids

Bibliography

Blaskovics ME, Nelson TL: Phenylketonuria and its variations: A review of recent developments. California Medicine 115:42–57, 1971

Hammond JE, Savory J: Advances in the detection of amino acids in biological fluids. Ann Clin Lab Sci 6:158–166, 1976

Holmgren G, Jeppson JO, Samuelson G: High-voltage electrophoresis in urinary amino acid screening. Scand J Clin Lab Invest 26: 313–318, 1970

Mass screening for amino acid disorders, editorial. Lancet 1:1133–1135, 1969

Newman RL, Starr DJ: Technology of a regional Guthrie test service. J Clin Pathol 24:564–575, 1971

Scriver CR, Clow CL, Lamm P: On the screening, diagnosis and investigation of hereditary aminoacidopathies. Clin Biochem 6: 142–188, 1973

Segal S: Disorders of renal amino acid transport. N Engl J Med 294: 1044–1051, 1976

Spiro RG: Glycoproteins: Their biochemistry, biology and role in human disease (first of two parts). N Engl J Med 281:991–1001, 1969

Stanbury JR, Wyngaarden JB (eds): The Metabolic Basis of Inherited Disease. New York: McGraw-Hill Book Co, 1978

Text (ASCP)

Dito WR, Patrick CW, Shelly J: Clinical Pathologic Correlations in Amniotic Fluid. Chicago: American Society of Clinical Pathologists, 1975 (Cat. no. 45-9-005-00)

Audiovisual Aid (ASCP)

Dito WR: Amniotic Fluid Analysis in Pregnancy at Risk. Chicago: American Society of Clinical Pathologists, 1975 (Cat. no. 47-9-001-00)

ELECTROLYTE, ACID-BASE, AND OXYGENATION DISTURBANCES

LEVEL I

Content

Pathophysiology—salt and water imbalance; acid-base disorders; disturbances in oxygenation

Diagnosis—interpretation of routine electrolyte/acid-base biochemical profiles; diagnostic test strategies for abnormal screening test results; indications for—and interpretation of—blood gases

Management—use of electrolyte and blood gas determinations in critically ill patients

Goals

The ability to interpret routine electrolyte/acid-base chemical profiles, and to devise strategies for further evaluation of sodium, potassium, and CO_2 abnormalities; the ability to interpret blood gas results in critically ill patients, and to advise on clinical management

Learning Aids

Bibliography

Acids, bases and nomograms, editorial. Lancet 2:814, 1974

Burke MD: Electrolyte studies. I. Sodium and water. Postgrad Med 64:147–153, 1978

Burke MD: Electrolyte studies. II. Potassium, chloride and acid-base. Postgrad Med 64:205–210, 1978

Comroe JH: Physiology of Respiration, ed 2. Chicago: Year Book Medical Publishers Inc, 1974

Harrington JT, Cohen JJ: Measurement of urinary electrolytes—indications and limitations. N Engl J Med 293:1241–1243, 1975

Rose BD: Introduction to disorders of osmolality, in Clinical Physiology of Acid-Base and Electrolyte Disorders. New York: McGraw-Hill Book Co, 1977

Schreier RW: Renal and Electrolyte Disorders. Boston: Little Brown & Co, 1976

Shapiro BA: Clinical Application of Blood Gases, ed 2. Chicago: Year Book Medical Publishers Inc, 1977

Snider GL: Interpretation of the arterial oxygen and carbon dioxide partial pressures. A simplified approach for bedside use. Chest 63: 801–806, 1973

Thomas HM, et al: The oxyhemoglobin dissociation curve in health and disease. Am J Med 57:331–348, 1974

Texts (ASCP)

Council on Clinical Chemistry

Baer DM: Acid-Base and Electrolyte Problems. Chicago: American Society of Clinical Pathologists, 1969 (Cat. no. 45-2-019-00—out of print)

Fleischer WR, Gambino SR: Blood pH, Po_2 and Oxygen Saturation. Chicago: American Society of Clinical Pathologists, 1972 (Cat. no. 45-2-029-00—out of print)

Audiovisual Aids (ASCP)

Fleischer WR, Hartmann AE: Blood Gases and Their Measurement—Part I. Chicago: American Society of Clinical Pathologists, 1976 (Cat. no. 47-2-022-00)

Fleischer WR, Hartmann AE: Blood Gases and Their Measurement—Part II. Chicago: American Society of Clinical Pathologists, 1976 (Cat. no. 47-2-024-00)

Fleischer WR, Weisberg H: Blood Gases and Pulmonary Function. Audiotape 38. Chicago: American Society of Clinical Pathologists, 1973 (Cat. no. 18-0-038-06—out of print)

LEVEL II

Content

Pathophysiology—advanced clinical-pathophysiologic case studies of electrolyte, acid-base, and oxygenation disturbances

Diagnosis—computerized approaches to the diagnosis and management of electrolyte and acid-base disorders

Goals

Sufficient understanding of the clinical and pathophysiologic aspects of electrolyte, acid-base, and oxygenation disturbances to solve complex clinical problems; a level of expertise in clinical interpretation surpassing that of an internist; ability to use computerized approaches in diagnosis and management

Learning Aids

Bibliography

Baylis PH, Heath DA: The development of a radioimmunoassay for the measurement of human plasma arginine vasopressin. Clin Endocrinol 7:91–102, 1977

Bleich HL: Computer-based consultation. Electrolyte and acid-base disorders. Am J Med 53:285–291, 1972

Burg MB: The nephron in transport of sodium, amino acids and glucose. Hospital Practice 13:99–109, 1978

Cannon PJ: The kidney in heart failure. N Engl J Med 296:26–32, 1977

Cohen JJ: Disorders of potassium balance. Hospital Practice 14: 119–128, 1979

Dormandy TL: Osmometry. Lancet 1:267–270, 1967

Emmett M, Narins RG: Clinical use of the anion gap. Medicine 56: 38–54, 1977

Feig PU, McCurdy DK: The hypertonic state. N Engl J Med 297: 1444–1454, 1977

Fulop M, Fulop M: Acid-base diagrams, maths, myths, and measurements. Lancet 2:637–639, 1974

Gennari FJ, Kassirer JP: Osmotic diuresis. N Engl J Med 291: 714–720, 1974

Hays RM: Principles of ion and water transport in the kidney. Hospital Practice 13:79–88, 1978

Kennedy PGE, Mitchell DM, Hoffbrand BI: Severe hyponatraemia in hospital inpatients. Br Med J 2:1251–1253, 1978

Levy M: The pathophysiology of sodium balance. Hospital Practice 13:95, 1978

Morris RC: Renal tubular acidosis. Mechanisms, classification and implications. N Engl J Med 281:1405–1413, 1969

Moses, AM: Diabetes insipidus and ADH regulation. Hospital Practice 12:37–44, 1977

Pontoppidan H, Geffin B, Lowenstein E: Acute respiratory failure in the adult. N Engl J Med 287:690–698, 743–752, 799–806, 1972

Vallbona C, Pevyn E, McMath F: Computer analysis of blood gases and of acid-base status. Comput Biomed Res 4:623–633, 1971

LEVEL III

Content

Pathophysiology—intracellular acid-base and electrolytes

Diagnosis—continuous monitoring techniques; the potential role of intracellular electrolytes and acid-base studies

Goals

An in-depth knowledge of acid-base chemistry; familiarity with continuous monitoring techniques and with the potential uses of intracellular electrolytes and acid-base studies

Learning Aids

Bibliography

Armstrong RF, et al: Continuous monitoring of mixed venous oxygen tension. Br Med J 2:282, 1976

Cohen PJ: The metabolic function of oxygen and biochemical lesions of hypoxia. Anesthesiology 37:148-177, 1972

Cox, M, Sterns RH, Singer I: The defense against hyperkalemia: The roles of insulin and aldosterone. N Engl J Med 299:525-532, 1978

Edelman IS, Leibman J: Anatomy of body water and electrolytes. Am J Med 27:256-277, 1959

Feeley TW, Hedley-Whyte J: Weaning from controlled ventilation and supplemental oxygen. N Engl J Med 292:903-906, 1975

Friis-Hansen B: Transcutaneous measurement of arterial blood oxygen tension with a new electrode. Scand J Clin Lab Invest 37:31-36, 1977

Hoffman JF: Ionic transport across the plasma membrane. Hospital Practice 9:119-127, 1974

Huch A, et al: Transcutaneous PCO_2 measurement with a miniaturised electrode. Lancet 1:982-983, 1977

Miller GA: Fat embolism: A comprehensive review. J Oral Surg 33: 91-103, 1975

Petty TL, Baily D, Best C: A new device for arterial blood gas sampling. JAMA 239:2016-2017, 1978

Plum F, Price RW: Acid-base balance of cisternal and lumbar cerebrospinal fluid in hospital patients. N Engl J Med 289:1346-1351, 1973

Severinghaus, JW: Acid-base balance nomogram—a Boston-Copenhagen detente. Anesthesiology 45:539-541, 1976

Vesterager P: Transcutaneous PO_2 electrode. Scand J Clin Lab Invest 37:27-30, 1977

RENAL DISEASE

LEVEL I

Content

Pathophysiology—glomerular dysfunction; tubular dysfunction

Diagnosis—urinalysis: interpretation of routine qualitative tests; diagnostic value of urinary sediment; strategy for diagnostic investigation of proteinuria; diagnosis of acute and chronic renal failure

Management—the role of laboratory tests in the management of acute and chronic renal failure

Goals

The ability to interpret results of routine urinalysis; to perform appropriate laboratory tests for acute and chronic renal failure in the proper sequence; and to advise on patient management

Learning Aids

Bibliography

Baek SM, et al: Free-water clearance patterns of predictors and therapeutic guides in acute renal failure. Surgery 77:632–640, 1975

Doolan PD, Alpen EL, Theil GB: A clinical appraisal of the plasma concentration and endogenous clearance of creatinine. Am J Med 32:65–79, 1962

Kerr DNS, Davison JM: The assessment of renal function. British Journal of Hospital Medicine 14:360–371, 1975

Levinsky NG: Pathophysiology of acute renal failure. N Engl J Med 296:1453–1458, 1977

Morgan DB, Dillon S, Payne RB: The assessment of glomerular filtration: Creatinine clearance or plasma creatinine? Postgrad Med J 54: 302–310, 1978

Renkin, EM, Robinson, RR: Glomerular filtration. N Engl J Med 290: 785–792, 1974

Schumann GB, Greenberg NF: Microscopic look at urine often unnecessary. JAMA 239:13–14, 1978

Atlases (ASCP)

Haber MH: Urine Casts: Their Microscopy and Clinical Significance. Chicago: American Society of Clinical Pathologists, 1976 (Cat. no. 15-9-002-00)

Lancaster RG, et al: Urinary Sediment. Formed Elements and Casts, Lancaster RG (ed). Chicago: American Society of Clinical Pathologists, 1970 (Cat. no. 15-9-001-00–out of print)

Audiovisual Aid (ASCP)

Haber MH: Urinalysis: Its Use in Clinical Diagnosis. Chicago: American Society of Clinical Pathologists, 1978 (Cat. no. 47-9-026-00)

LEVEL II

Content

Pathophysiology–correlation of ultrastructure and function in the pathophysiology of renal disease

Diagnosis–the role of immunochemical and immunofluorescent techniques in the diagnosis of renal disease

Management–decision analysis in the diagnosis and management of acute and chronic renal failure

Goals

In-depth understanding of renal physiology and renal disease, of the application and interpretation of immunochemical and immunofluorescent techniques, and of decision analysis in diagnosis and management

Learning Aids

Bibliography

Baldwin DS: Poststreptococcal glomerulonephritis. A progressive disease? Am J Med 62:1–11, 1977

Baldwin DS, et al: Clinical course as related to morphologic forms and their transitions. Am J Med 62:12–30, 1977

Brenner BM, Hostetter TH, Humes HD: Molecular basis of proteinuria of glomerular origin. N Engl J Med 298:826–833, 1978

Gorry GA, et al: Decision analysis as the basis for computer-aided management of acute renal failure. Am J Med 55:473–484, 1973

Gyory AZ, et al: Comprehensive one-day renal function testing in man. J Clin Pathol 27:382–391, 1974

Pollak VE: Proteinuria. I. Mechanisms. II. Diagnosis and management. Hospital Practice 6:49–56, 6:59–73, 1971

Spargo BH: Practical use of electron microscopy for the diagnosis of glomerular disease. Hum Pathol 6:405–420, 1975

Wills MR: Biochemical consequences of chronic renal failure: A review. J Clin Pathol 21:541–554, 1968

Text and Atlas (ASCP)

Pirani CL, McCluskey RT: Medical Diseases of the Kidney. Chicago: American Society of Clinical Pathologists, 1972 (Proceedings: Cat. no. 50-1-038-00/Atlas and slides: Cat. no. 15-1-012-00)

Audiovisual Aids (ASCP)

Smith RD, Weiss MA: Renal Biopsy: Technical Aspects of Immunofluorescence and Electron Microscopy. Chicago: American Society of Clinical Pathologists, 1979 (Cat. no. 47-1-033-00)

Smith RD, Weiss MA: Renal Biopsy: Technical Aspects of Light Microscopic Interpretation. Chicago: American Society of Clinical Pathologists, 1979 (Cat. no. 47-1-032-00)

HEPATIC DISEASE

LEVEL I

Content

Pathophysiology—the concept of organelle dysfunction in relation to hepatic function tests

Diagnosis—test strategies for the detection, specific pathologic findings, and management of hepatic disease; the construction of liver profiles and the intralaboratory strategies for their interpretation

Goals

The ability to interpret liver profiles, and to advise on the choice and sequence of test combinations for specific diagnosis and management of hepatic disease

Learning Aids

Bibliography

Boone DJ, Routh JI, Schrantz R: Gamma-glutamyl transpeptidase and 5′-nucleotidase. Comparison as diagnostics for hepatic disease. Am J Clin Pathol 61:321–327, 1974

Burke MD: Hepatic function testing. Postgrad Med 64:177–185, 1978

Burke MD: Liver function. Hum Pathol 6:273–286, 1975

Ewen LM: Separation of alkaline phosphatase isoenzymes and evaluation of the clinical usefulness of this determination. Am J Clin Pathol 61:142–154, 1974

Gitnick GL: Viral hepatitis. West J Med 128:117–126, 1978

Kaplowitz N: Cholestatic liver disease. Hospital Practice 13:83–92, 1978

Klatskin G, Kantor FS: Mitochondrial antibody in primary biliary cirrhosis and other diseases. Ann Intern Med 77:533–541, 1972

Leevy CM, Kanagasundaram N: Alcoholic hepatitis. Hospital Practice 13:115–123, 1978

Mihas AA, Conrad ME: Hepatitis B antigen and the liver. Medicine 57:129–150, 1978

Sherlock S: Primary biliary cirrhosis. Am J Med 65:217–219, 1978

Whitehead TP, Clarke CA, Whitfield G,: Biochemical and haematological markers of alcoholic intake. Lancet 1:978–981, 1978

Texts (ASCP)

Council on Clinical Chemistry

Batsakis JG, Briere RO, Markel SF: Diagnostic Enzymology. Chicago: American Society of Clinical Pathologists, 1972, pp 19–49 (Cat. no. 45-2-026-00—out of print)

Fuller JB (ed): Liver Function Tests. Chicago: American Society of Clinical Pathologists, 1966 (Cat. no. 45-2-010-00—out of print)

LEVEL II

Content

Discriminant function analysis in the selection of appropriate profiles for the diagnosis, specific pathology and management of disease; multivariate analysis of test results

Goals

Ability to apply quantitative techniques to the interpretation of multiple hepatic function tests

Learning Aids

Bibliography

Baron DH: A critical look at the value of biochemical liver function tests with special reference to discriminant function analysis. Ann Clin Biochem 7:100–103, 1970

Baron DN, Path FC, Fraser PM: Medical applications of taxonomic methods. Br Med Bull 24:236–240, 1968

Fraser PM, Franklin DA: Mathematical models for the diagnosis of liver disease. Q J Med 43:73–88, 1974

Sher PP: Diagnostic effectiveness of biochemical liver function tests, as evaluated by discriminant function analysis. Clin Chem 23: 627–630, 1977

Winkel P, et al: Diagnostic value of routine liver tests. Clin Chem 21: 71–75, 1975

Audiovisual Aids (ASCP)

Peters RL: Chronic Hepatitis. Chicago: American Society of Clinical Pathologists, 1978 (Cat. no. 47-1-029-00)

Peters RL: Liver Biopsy: An Approach to Non-specific Findings. Chicago: American Society of Clinical Pathologists, 1975 (Cat. no. 47-1-007-00)

Peters RL: Morphology of Acute Viral Hepatitis. Chicago: American Society of Clinical Pathologists, 1976 (Cat. no. 47-1-016-00)

LEVEL III

Content

Pathophysiology—disorders of bilirubin metabolism, cholestasis; correlation of ultrastructure and function

Diagnosis—the potential value of intracellular biochemical analysis in the clinical evaluation of hepatic disease

Goals

Familiarity with the clinical consequences of hepatic pathophysiology, and with new approaches in cellular biochemistry; and clinical expertise in gastroenterology

Learning Aids

Bibliography

Barnes S, et al: Diagnostic value of serum bile acid estimations in liver disease. J Clin Pathol 28:506–509, 1975

Berk PD, et al: Unconjugated hyperbilirubinemia. Physiologic evaluation and experimental approaches to therapy. Ann Intern Med 82:552–570, 1975

Bissell DM: Formation and elimination of bilirubin. Gastroenterology 69:519–538, 1975

Bouchier IAD, Pennington CR: Serum bile acids in hepatobiliary disease. Gut 19:492–496, 1978

Brocklehurst D, Lathe GH, Aparicio SR: Serum alkaline phosphatase, nucleotide pyrophosphatase, 5′-nucleotidase and lipoprotein-X in cholestasis. Clin Chim Acta 67:269–279, 1976

Dienstag JL, Isselbacher KJ: Liverspecific protein: More questions than answers. N Engl J Med, 299:40–42, 1978

Dietrichson O, Christoffersen P: The prognosis of chronic aggressive hepatitis. A clinical and morphological follow-up study. Scand J Gastroenterology 12:289–295, 1977

Erlinger S: Cholestasis: Pump failure, microvilli defect, or both. Lancet 1:533–534, 1978

Javitt NB: Hepatic bile formation. N Engl J Med 295:1464–1469, 1511–1516, 1976

Korman MG, Hofmann AF, Summerskill WH: Assessment of activity in chronic active liver disease. Serum bile acids compared with conventional tests and histology. N Engl J Med 290:1399–1402, 1974

Lieber CS: Pathogenesis and early diagnosis of alcoholic liver injury. N Engl J Med 298:888–893, 1978

Magnani HN, Alaupovic P: Utilization of the quantitative assay of lipoprotein-X in the differential diagnosis of extrahepatic obstructive jaundice and intrahepatic diseases. Gastroenterology 71:87–93, 1976

Miller JP, Johnson K: A critical examination of the value of combined determinations of lecithin: Cholesterol acyltransferase and lipoprotein-X in the differential diagnosis of liver disease. Clin Chim Acta 69:81–84, 1976

Seymour CA, Peters TJ: Enzyme activities in human liver biopsies: Assay methods and activities of some lysosomal and membrane-bound enzymes in control tissue and serum. Clin Sci Mol Med 52: 229–239, 1977

Text (ASCP)

Council on Clinical Chemistry

Gambino SR, Di Re J: Bilirubin Assay. Chicago: American Society of Clinical Pathologists, 1968 (Cat. no. 45-2-023-00 – out of print)

CARDIOVASCULAR DISEASE

LEVEL I

Content

Pathophysiology – myocardial injury; clinical-pathologic correlation in angina, preinfarction angina, and acute myocardial infarction; hypertension, biochemical correlates

Diagnosis – test strategies for the diagnosis of acute myocardial infarction; test strategies for the diagnosis and management of the hypertensive patient

Goals

Sufficient knowledge of acute myocardial disease and hypertension to interpret enzyme test results, and to advise on investigation of hypertension

Learning Aids

Bibliography

Blomberg DJ, Kimber W, Burke MD: Creatine kinase isoenzymes. Predictive value in the early diagnosis of acute myocardial infarction. Am J Med 59:464–469, 1975

Burke MD: Clinical enzymology. 2. Test strategies and interpretation of results. Postgrad Med 64:149–156, 1978

Burke MD: Hypertension: Exploring the great unknown. Diagnostic Medicine 1:34–41, 1978

Burke MD: Hypertension: Test strategies for diagnosis and management. Diagnostic Medicine 2:72–84, 1979

Galen RS, Reiffel JA, Gambino SR: Diagnosis of acute myocardial infarction. Relative efficiency of serum enzyme and isoenzyme measurêments. JAMA 232:145–147, 1975

Gitlow SE, Mendlowitz M, Bertani LM: The biochemical techniques for detecting and establishing the presence of a pheochromocytoma. Am J Cardiol 26:270–279, 1970

Goldberg DM, Winfield DA: Diagnostic accuracy of serum enzyme assays for myocardial infarction in a general hospital population. Br Heart J 34:597–604, 1972

Hillis LD, Braunwald E: Myocardial ischemia. N Engl J Med 296: 971–978, 296:1034–1041, 296:1093–1096, 1977

Horton R: Aldosterone: Review of its physiology and diagnostic aspects of primary aldosteronism. Metabolism 22:1525–1545, 1973

Roe CR, et al: Relation of creatine kinase isoenzyme MB to postoperative electrocardiographic diagnosis in patients undergoing coronary-artery bypass surgery. Clin Chem 25:93–98, 1979

Shine KE, et al: Pathophysiology of myocardial infarction. Ann Intern Med 87:75–85, 1977

Swales JD: The hunt for renal hypertension. Lancet 1:577–579, 1976

Texts (ASCP)

Council on Clinical Chemistry

Batsakis JG, Briere RO, Markel SF: Diagnostic Enzymology. Chicago: American Society of Clinical Pathologists, 1972 (Cat. no. 45-2-026-00—out of print)

Other Councils

Budinger JM (ed): Cardiovascular Pathology. Chicago: American Society of Clinical Pathologists, 1978 (Cat. no. 50-1-042-00)

LEVEL II

Content

Pathophysiology—atherosclerosis: biochemical correlates

Diagnosis and management—quantitation of infarct size, and techniques for modifying acute myocardial infarct; critical evaluation of the role of renin profiling in essential hypertension

Goals

Knowledge of the biochemistry of atherosclerosis, of the clinical pathophysiology of acute myocardial infarction, and of the methods for quantitating infarct size. The ability to evaluate the use of renin profiling in the work-up of essential hypertension

Learning Aids

Bibliography

Kaplan NM: Renin profiles. The unfulfilled promises. JAMA 238: 611–613, 1977

Parisi AF, et al: Noninvasive cardiac diagnosis. N Engl J Med 296: 315–320, 296:368–374, 296:427–432, 1977

Pickering G: Arteriosclerosis and atherosclerosis. Am J Med 34:7–18, 1963

Roe CR, Cobb FR, Starmer, CF: The relationship between enzymatic and histologic estimates of the extent of myocardial infarction in conscious dogs with permanent coronary occlusion. Circulation 55: 438–449, 1977

Ross R, Glomset JA: The pathogenesis of atherosclerosis. N Engl J Med 295:369–377, 295:420–425, 1976

Sobel BE, Roberts R, Larson KB: Considerations in the use of biochemical markers of ischemic injury. Circ Res 38(suppl 1): I99–I108, 1976

Sobel BE, Roberts R, Larson KB: Estimation of infarct size from serum MB creatine phosphokinase activity: Applications and limitations. Am J Cardiol 37:474–485, 1976

Stason WB, Weinstein MC: Allocation of resources to manage hypertension. N Engl J Med 296:732–739, 1977

Varki AP, Roby DS, Watts H: Serum myoglobin in acute myocardial infarction: A clinical study and review of the literature. Am Heart J 96:680–688, 1978

Waters DD, Forrester JS: Myocardial ischemia: Detection and quantitation. Ann Intern Med 88:239–250, 1978

ENDOCRINE DISEASES

LEVEL I

Content

Thyroid Disease

Pathophysiology—hypothalamic-pituitary-thyroid interrelationships; autonomous hyperthyroidism; primary vs secondary hypothyroidism

Diagnosis—strategies for the detection and diagnosis of thyroid dysfunction

Management—laboratory strategy in the management of treated hypothyroid and hyperthyroid states

Adrenocortical Disease

Pathophysiology—normal and disordered adrenocortical steroid biosynthesis; hypothalamic-pituitary-adrenocortical interrelationships; adrenocortical dysfunctional states; adrenal insufficiency

Diagnosis—test strategies for the detection of Cushing's syndrome, congenital adrenal hyperplasia and adrenocortical insufficiency; test strategies for the specific etiologic diagnosis of Cushing's syndrome, congenital adrenal hyperplasia and adrenal virilism

Management—laboratory management of steroid withdrawal

Parathyroid Disease

Pathophysiology—altered calcium ion homeostasis, the role of parathyroid hormone (PTH), vitamin D metabolites, calcitonin and PO_4; autonomous vs homeostatic hyperparathyroidism; hypoparathyroidism and pseudohypoparathyroidism

Diagnosis—differential diagnosis of hypercalcemia; test strategy for the diagnosis of primary hyperparathyroidism; screening and interpretation of profiles for the detection and diagnosis of metabolic bone diseases

Pregnancy and the Fetoplacental Unit

Pathophysiology—biochemical alterations in pregnancy; normal and abnormal development of the fetoplacental unit

Diagnosis—test strategies for the assessment of fetoplacental unit; indications for amniocentesis in late pregnancy

Management—routine laboratory management of the pregnant patient; prediction and management of eclamptic states

Gonadal Disease

Pathophysiology—hypothalamic-pituitary-gonadal interrelationships; hypogonadism; precocious pubertal states; ovarian vs adrenal virilism; feminizing states

Diagnosis—test strategies for the laboratory evaluation of amenorrhea, infertility, sexual dysfunction, precocious puberty, and virilizing and feminizing states

Hypothalamic-pituitary Disease

Pathophysiology—hypothalamic facilitation and inhibition; the sequence of pituitary trophic hormone loss in pituitary disease; growth hormone excess

Diagnosis—test strategies for the evaluation of hypothalamic-pituitary dysfunction; the assessment of pituitary reserve

Goals

Comprehension of the feedback arrangement between hypothalamic-pituitary levels and target organs; and appreciation of the difference between homeostatic and autonomous endocrine abnormalities. Understanding of the chemistry of steroids and steroid biosynthesis in the adrenal cortex and gonads; of the role of vitamin D in altered calcium ion homeostasis; and of the pathophysiology of metabolic bone disease. Ability to offer internists consultative advice on test strategies for detection, specific diagnosis, and management of endocrine dysfunction.

Learning Aids

Bibliography

Thyroid Disease

Abuid J, Larsen PR: Triiodothyronine and thyroxine in hyperthyroidism. Comparison of the acute changes during therapy with antithyroid agents. J Clin Invest 54:201–208, 1974

Britton KE, et al: A strategy for thyroid function tests. Br Med J 3: 350–352, 1975

Irvine WJ, Toft AD: The diagnosis and treatment of thyrotoxicosis. Clin Endocrinol 5:687–707, 1976

Steffes MW: Testing for hypothyroidism. Lab 78 1:9–14, 1978

Steffes MW: Testing for hyperthyroidism. Lab 79 2:18–23, 1979

Stock JM, Surks MI, Oppenheimer JH: Replacement dosage of L-thyroxine in hypothyrodism. A re-evaluation. N Engl J Med 290: 529–533, 1974

Adrenocortical Disease

Broughton A: Application of adrenocorticotropin assays in a routine clinical laboratory, Am J Clin Pathol 64:618–624, 1975

Byyny RL: Withdrawal from glucocorticoid therapy. N Engl J Med 295:30–32, 1976

Hankin ME, Theile HM, Steinbeck AW: An evaluation of laboratory tests for the detection and differential diagnosis of Cushing's syndrome. Clin Endocrinol 6:185–196, 1977

Hughes IA, Winter JS: The application of a serum 17OH-progesterone radioimmunoassay to the diagnosis and management of congenital adrenal hyperplasia. J Pediatr 88:766–773, 1976

Jenkins JS: The assessment of adrenal function. British Journal of Hospital Medicine 14:373–380, 1975

Migeon CJ: Adrenal androgens in man. Am J Med 53:606–626, 1972

Migeon CJ: Diagnosis and management of congenital adrenal hyperplasia. Hospital Practice 12:75–82, 1977

Nichols T, Nugent CA, Tyler FH: Steroid laboratory tests and the diagnosis of Cushing's syndrome. Am J Med 45:116–128, 1968

Parathyroid Disease

De Luca HF: Vitamin D endocrinology. Ann Intern Med 85:367–377, 1976

Lee DB, Zawada ET, Kleeman CR: The pathophysiology and clinical aspects of hypercalcemic disorders. West J Med 129:278–320, 1978

Lutwak L, Singer FR, Urist MR: Current concepts of bone metabolism. Ann Intern Med 80:630–644, 1974

Posen S, et al: Parathyroid hormone assay in clinical decision-making. Br Med J 1:16–19, 1976

Purnell DC, Scholz DA, Smith LH: Diagnosis of primary hyperparathyroidism. Surg Clin North Am 57:543-556, 1977

Queener SF, Bell NH: Calcitonin: A general survey. Metabolism 24: 555–567, 1975

Rasmussen H, et al: Hormonal control of skeletal and mineral homeostasis. Am J Med 56:751–758, 1974

Reiss E, Canterbury JM: Blood levels of parathyroid hormone in disorders of calcium metabolism. Annu Rev Med 24:217–232, 1973

Pregnancy and the Fetoplacental Unit

Chamberlain G: Predicting and evaluating fetal distress in labor. Postgrad Med 59:151–158, 1976

Dewhurst CJ: Recognizing the fetus at risk. Postgrad Med 59: 114–117, 1976

Dickey RP, Grannis CF, Hanson FW: Use of the estrogen-creatinine ratio and the "estrogen index" for screening of normal and "high-risk" pregnancy. Am J Obstet Gynecol 113:880–886, 1972

Klopper A: Monitoring the fetoplacental unit—the fetus. Postgrad Med J 51:227–230, 1975

Lind T: The assessment of normal pregnancy. British Journal of Hospital Medicine 14:253–256, 1975

Predicting fetal death, editorial. Br Med J 1:123–124, 1977

Talbert LM, Easterling WE Jr:Factors influencing urinary estrogen excretion in pregnancy. Am J Obstet Gynecol 99:923–932, 1967

Tulchinsky D, Osathanondh R, Finn A: Dehydroepiandrosterone sulfate loading test in the diagnosis of complicated pregnancies. N Engl J Med 294:517–522, 1976

Whitfield CR: The diagnostic value of amniocentesis. Clin Obstet Gynaecol 1:67–84, 1974

Wilde CE, Oakey RE: Biochemical tests for the assessment of fetoplacental function. Ann Clin Biochem 12:83–118, 1975

Gonadal Disease

De Kretser DM: The management of the infertile male. Clin Obstet Gynaecol 1:409–427, 1974

Federman DD: Disorders of sexual development. N Engl J Med 277: 351–360, 1967

Gilson MD, Knab DR: Primary amenorrhea: A simplified approach to diagnosis. Am J Obstet Gynecol 117:400–406, 1973

Greenblatt RB: Diagnosis and treatment of hirsutism. Hospital Practice 8:91–98, 1973

Lipsett M, Wessler S, Avioli LV: The differential diagnosis of hirsutism and virilism. Arch Intern Med 132:616–619, 1973

London DR: Medical aspects of hypogonadism. Clin Endocrinol Metabol 4:597–618, 1975

Paulsen CA: Recognition and management of testicular failure. Hospital Practice 7:133–142, 1972

Posalaky Z: Semen analysis and male infertility. Minn Med 59: 877–886, 1976

Rifka SM, Sherins RJ: Current concepts in the evaluation of the infertile male. Clin Obstet Gynaecol 5:481–497, 1978

Hypothalamic-pituitary Disease

Catt KJ: II. Pituitary function. Lancet 1:827–831, 1970

Catt KJ: III. Growth hormone. Lancet 1:933–939, 1970

Gold EM: Hypothalamic-pituitary function tests. Postgrad Med 62: 105–114, 1977

Ontjes DA, Ney RL: Tests of anterior pituitary function. Metabolism 21:159–177, 1972

Spark RF: Simplified assessment of pituitary-adrenal reserve. Measurement of serum 11-deoxycortisol and cortisol after metapyrone. Ann Intern Med 75:717–723, 1971

Texts (ASCP)

Council on Clinical Chemistry

Biskind GR: Hormone Assay Technical Manual. Chicago: American Society of Clinical Pathologists, 1965 (Cat. no. 45-2-008-00—out of print)

Bittner DL, Gambino SR: Thyroid Disease: PBI, T4, Free Thyroxine or T3? Chicago: American Society of Clinical Pathologists, 1968 (Cat. no. 45-2-024-00—out of print)

Clerch AR: Adrenal Cortex Function. Chicago: American Society of Clinical Pathologists, 1972 (Cat. no. 45-2-027-00—out of print)

Other Councils (ASCP)

Dito WR, Patrick CW, Shelly J: Clinical Pathologic Correlations in Amniotic Fluid. Chicago: American Society of Clinical Pathologists, 1975 (Cat. no. 45-9-005-00)

Woodward SC, Hansell JR, Bering NM: The Determinations of T-3 and T-4. Chicago: American Society of Clinical Pathologists, 1973 (Cat. no. 45-8-005-00—out of print)

Audiovisual Aids (ASCP)

Dito WR: Amniotic Fluid Analysis in Pregnancy at Risk. Chicago: American Society of Clinical Pathologists, 1975 (Cat. no. 47-9-001-00)

LEVEL II

Content

Pathophysiology—disorders of thyroid hormone biosynthesis; T_3-T_4 relationships and the relevance of reverse T_3; vitamin D metabolism; biochemical aspects of metabolic bone disease;

metabolic effects of oral contraception; disorders of sex differentiation; neuroendocrinology; gastrointestinal hormones; the prostaglandins

Diagnosis—discriminant function analysis in selection of appropriate profiles for investigation of endocrinologic disease; multivariate analysis test results; test strategies for diagnosis of Zollinger-Ellison syndrome and of pancreatic diarrheal disorders

Goals

Knowledge of gastrointestinal endocrinology; expertise in the pathophysiology of endocrine disease to aid in the solution of difficult and complex clinical problems; and familiarity with multivariate analysis of test results in endocrinologic disease

Learning Aids

Bibliography

Archer DF: Current concepts of prolactin physiology in normal and abnormal conditions. Fertil Steril 28:125–134, 1977

Besser GM, Mortimer CH: Hypothalamic regulatory hormones: A review. J Clin Pathol 27:173–184, 1974

Braverman LE, Ingbar SH, Sterling K: Conversion of thyroxine (T_4) to triiodothyronine (T_3) in athyreotic human subjects. J Clin Invest 49: 855–864, 1970

Britton KE, Quinn V, Ellis SM: Is "T_4 toxicosis" a normal biochemical finding in elderly women? Lancet 2:141–142, 1975

Brown J, et al: Autoimmune thyroid diseases—Graves' and Hashimoto's. Ann Intern Med 88:379–391, 1978

Burman KD: Recent developments in thyroid hormone metabolism: Interpretation and significance of measurements of reverse T_3, 3, 3′, T_2, and thyroglobulin. Metabolism 27:615–630, 1978

Burman KD, et al: A radioimmunoassay for 3, 3′, 5′-L-triiodothyronine (reverse T_3): Assessment of thyroid gland content and serum measurements in conditions of normal and altered thyroidal economy and following administration of thyrotropin releasing hormone

(TRH) and thyrotropin (TSH). J Clin Endocrinol Metab 44: 660–672, 1977

Catt KJ: IV. Reproductive endocrinology. Lancet 1:1097–1104, 1970

Crawford JD: It's a boy? N Engl J Med 291:976–977, 1974

Fisher DA: Screening for congenital hypothyroidism. Hospital Practice 12:73–78, 1977

Harrison RF, Roberts AP, Campbell S: A critical evaluation of tests used to assess gestational age. Br J Obstet Gynaecol 84:98–107, 1977

Harsoulis P, et al: Combined test for assessment of anterior pituitary function. Br Med J 4:326–329, 1973

Larsen PR: Thyroidal triiodothyronine and thyrocine in Graves' disease: Correlation with presurgical treatment, thyroid status, and iodine content. J Clin Endocrinol Metab 41:1098–1104, 1975

McGirr EM: Inherited defects in thyroid hormone synthesis. Ann Clin Res 4:200–203, 1972

Montgomery, DAD, Harley JMG: Endocrine disorders: Disorders of the pituitary gland. Clin Obstet Gynaecol 4:339–370, 1977

Nugent, CA, et al: Probability theory in the diagnosis of Cushing's syndrome. J Clin Endocrinol 24:621–627, 1964

Refetoff S, et al: Metabolism of thyroxine-binding globulin in man. Abnormal rate of synthesis in inherited thyroxine-binding globulin deficiency and excess. J Clin Invest 57:485–495, 1976

Segal S, Polishuk WZ, Ben-David M: Hyperprolactinemic male infertility. Fertil Steril 27:1425–1427, 1976

Squire CR, Gimlette TM: Comparison of thyroid stimulating hormone and triiodothyronine response to thyrotrophin releasing hormone in the assessment of thyroid status. J Clin Pathol 30:635–637, 1977

Starling JR, Harris C, Granner DK: Diagnosis of occult familial medullary carcinoma of the thyroid using pentagastrin. Arch Surg 113: 241–243, 1978

Thorner MO: Prolactin. Clin Endocrinol Metabol 6:201–222, 1977

Text and Atlas (ASCP)

Hartmann WH, Warner NE, Oertel JE: Endocrine Pathology. Chicago: American Society of Clinical Pathologists, 1978 (Cat. no. 50-1-043-00)

Mostofi FK, Scully RE: Lesions of the Gonads, Gall EA (ed). Chicago: American Society of Clinical Pathologists, 1973 (Cat. no. 15-1-011-00—out of print)

LEVEL III

Content

Pathophysiology—hormone action and cellular control mechanisms; genetics and endocrinology; hormonal aspects of growth, development, and aging

Diagnosis and management—new developments in the diagnosis and management of neuroendocrine disease; cellular biochemistry in diagnosis and management

Goals

Expertise in all areas of endocrinology, including familiarity with new developments in diagnosis and management of endocrine disease

Learning Aids

Bibliography

Brown GM: Psychiatric and neurologic aspects of endocrine disease. Hospital Practice 10:71–79, 1975

Catt KJ: I. Hormones in general. Lancet 1:763–765, 1970

Jewelewicz R: The diagnosis and treatment of amenorrheas. Fertil Steril 27:1347–1358, 1976

McEwen BS: The brain as a target organ of endocrine hormones. Hospital Practice 10:95–104, 1975

Peck WA: Comments on the importance of some recent neuroendocrinology advances. Arch Intern Med 135:1362–1363, 1975

Rimoin DL: Genetic defects of growth hormone. Hospital Practice 6: 113–124, 1971

Smythe GA: The role of serotonin and dopamine in hypothalamic-pituitary function. Clin Endocrinol 7:325–341, 1977

Sterling K: Thyroid hormone action at the cell level. N Engl J Med 300:117–123, 173–177, 1979

Tischler AS, et al: Neuroendocrine neoplasms and their cells of origin. N Engl J Med 296:919–925, 1977

CHEMICAL HEMATOLOGY

LEVEL I

Content

Pathophysiology—heme proteins; hemoglobinopathy; thalassemias; abnormal hemoglobin derivatives; iron metabolism

Diagnosis—test strategies for the investigation of iron metabolism disorders; routine investigation of the hemoglobinopathies

Goals

Sufficient command of biochemical methods in routine hematology to devise test strategies for disorders of iron metabolism and to interpret results; familiarity with the routine investigation of the hemoglobinopathies

Learning Aids

Bibliography

Davis CS: Diagnostic value of muramidase. Postgrad Med 49:51–54, 1971

Hershko C: The fate of circulating haemoglobin. Br J Haematol 29: 199–204, 1975

Hussein S, et al: Serum ferritin in megaloblastic anemia. Scand J Haematol 20:241–245, 1978

Jacobs A, Worwood M: The clinical use of serum ferritin estimation. Br J Haematol 31:1–3, 1975

Jacobs A, Worwood M: Ferritin in serum. Clinical and biochemical implications. N Engl J Med 292:951–956, 1975

Jenkins DT, Wishart MM, Schenberg C: Serum ferritin in pregnancy. Aust NZ J Obstet Gynaecol 18:223–225, 1978

Osserman EF, Lawlor DP: Serum and urinary lysozyme (muramidase) in monocytic and monomyelocytic leukemia. J Exp Med 124: 921–952, 1966

Schmidt RM: Laboratory diagnosis of hemoglobinopathies. JAMA 224:1276–1280, 1973

Smith RP, Olson MV: Drug-induced methemoglobinemia. Semin Hematol 10:253–268, 1973

Weatherall DJ, Clegg JB: Hereditary persistence of fetal haemoglobin. Br J Haematol 29:191–198, 1975

Wilkinson JH: Plasma enzymes in anemia, in The Principles and Practice of Diagnostic Enzymology. Chicago: Year Book Medical Publishers Inc, 1976

Audiovisual Aids (ASCP)

Dutcher TF: Bone Marrow Interpretation in Selected Anemias. Chicago: American Society of Clinical Pathologists, 1976 (Cat. no. 47-5-020-00)

Dutcher TF: Introduction to the Interpretation of Bone Marrow Sections. Chicago: American Society of Clinical Pathologists, 1975 (Cat. no. 47-5-002-00)

Hoffman GC: Hemoglobinopathies. Chicago: American Society of Clinical Pathologists, 1972 (Cat. no. 21-5-011-00—out of print)

Koepke JA: Iron Deficiency Anemia. Chicago: American Society of Clinical Pathologists, 1974 (Cat. no. 21-5-012-00—out of print)

Miale JB: Megaloblastic Anemias. Chicago: American Society of Clinical Pathologists, 1966 (Cat. no. 21-5-006-00—out of print)

LEVEL II

Content

Pathophysiology—inborn errors of red cell metabolism; abnormal derivatives of hemoglobin

Diagnosis—test strategies for specific diagnosis of hemoglobin variants, thalassemias, abnormal hemoglobin derivatives, and red cell enzyme defects

Goals

Knowledge of inborn errors of red cell metabolism and of abnormal derivatives of hemoglobin; ability to conduct test strategies for the specific diagnosis of hemoglobin variants, thalassemias, abnormal hemoglobin derivatives, and red cell enzyme defects

Learning Aids

Bibliography

Beutler E: Enzyme tests in hematological diseases, in Wilkinson JH (ed): in The Principles and Practice of Diagnostic Enzymology. Chicago: Year Book Medical Publishers Inc, 1976, pp 423–460

Geokas MC, et al: Methemalbumin in the diagnosis of acute hemorrhagic pancreatitis. Ann Intern Med 81:483–486, 1974

McCurdy PR: Clinical manifestations of variant G-6-PD enzymes. Ann Clin Lab Sci 1:184–192, 1971

Morse EE, Jilani F, Brassel J: Acquired pyruvate kinase deficiency. Ann Clin Lab Sci 7:399–404, 1977

Pearson HA, et al: Comprehensive testing for thalassemia trait. Ann N.Y Acad Sci 232:135–143, 1974

Stockman JA, Weiner LS, Simon GE: The measurement of free erythrocyte porphyrin (FEP) as a simple means of distinguishing iron deficiency from beta-thalassemia trait in subjects with microcytosis. J Lab Clin Med 85:113–119, 1975

Waxman S: Metabolic approach to the diagnosis of megaloblastic anemias. Med Clin North Am 37:315–334, 1973

Yoshida A: Hemolytic anemia and G6PD deficiency. Science 179: 532–537, 1973

DISEASES OF SKELETAL MUSCLE

LEVEL I

Content

Pathophysiology—myopathy and muscle injury; serum enzyme activity and muscle injury

Diagnosis—clinical enzymology and muscle disease; test strategies for the detection and differential diagnosis of myopathy

Management—laboratory tests in the management of muscular disease

Goals

Basic familiarity with myopathy, muscle injury, and with the clinical enzymology of muscle disease; the ability to conduct test strategies for the detection and differential diagnosis of myopathy; and familiarity with laboratory tests in the management of muscular diseases

Learning Aids

Bibliography

Brownlow K, Elevitch FR: Serum creatine phosphokinase isoenzyme (CPK_2) in myositis. A report of 6 cases. JAMA 230:1141–1144, 1974

Furukawa T, Peter JB: The muscular dystrophies and related disorders. I. The muscular dystrophies. JAMA 239:1537–1542, 1978

Munsat TL, et al: Serum enzyme alterations in neuromuscular disorders. JAMA 226:1536–1543, 1973

Somer H, Donner M, Murros J: A serum isoenzyme study in muscular dystrophy. Arch Neurol 29:343–345, 1973

Thomson WHS: Serum enzyme studies in inherited disease of muscle. Clin Chim Acta 35:183–191, 1971

Vignos PJ Jr, Goldwyn J: Evaluation of laboratory tests in diagnosis and management of polymyositis. Am J Med Sci 263:291–308, 1972

Vladutiu AO, Venuto RC: Creatine kinase MB and lactate dehydrogenase. 5 isoenzymes in rhabdomyolysis. Clin Chem 23:1366, 1977

Text (ASCP)

Council on Clinical Chemistry

Batsakis JG, Briere RO, Markel SF: Diagnostic Enzymology. Chicago: American Society of Clinical Pathologists, 1972 (Cat. no. 45-2-026-00–out of print)

LEVEL II

Content

Pathophysiology–correlation of structure and function at the neuromuscular junction and the muscle cell in reaction to injury; regeneration, atrophy, hypertrophy; disorders of glycogenolysis, aerobic metabolism, cell growth and development, contractile elements

Diagnosis and management–discriminant function analysis and multivariate analysis in the choice of test profiles and the interpretation of results for diagnosis and management of muscle diseases; detection of carriers in muscular dystrophies; detection of susceptibility to malignant hyperpyrexia

Goals

A more in-depth understanding of structure and function in neuromuscular and muscle injury and of statistical approaches and decision analysis in diagnosis of muscle disease. Familiarity with methods for detection of carriers in the muscular dystrophies and with methods for detection of susceptibility to malignant hyperpyrexia.

Learning Aids

Bibliography

Association for Research in Nervous and Mental Disease: Neuromuscular abnormalities in the major mental illnesses. I. Serum enzyme studies, in Freedman DX (ed): Biology of the Major Psychoses: A Comparative Analysis. New York: Raven Press, 1975, vol 54

Hutton EM, Thompson MW: Carrier detection and genetic counseling in Duchenne muscular dystrophy: a follow-up study. Can Med Assoc J 115:749, 1976

Thorstensson A: Muscle strength, fibre types and enzyme activities in man. Acta Physiol Scand (suppl 443):1–45, 1976

Zellweger H, Antonik A: Newborn screening for Duchenne muscular dystrophy. Pediatrics 55:30–34, 1975

GASTROINTESTINAL AND PANCREATIC DISEASES

LEVEL I

Content

Pathophysiology—gastric dysfunction; intestinal malabsorption; bile salt metabolism

Diagnosis—test strategies for the detection of acid-peptic disease and the Zollinger-Ellison syndrome; diagnosis of acute pancreatitis; test strategies for the diagnosis of intestinal malabsorption and chronic pancreatic disease

Goals

Sufficient knowledge of the pathophysiology of acid-peptic disease and malabsorption to develop a test strategy for intestinal malabsorption and to interpret the results of gastric acid analyses; familiarity with an optimal test strategy for the diagnosis of acute pancreatitis

Learning Aids

Bibliography

Baron JH: The clinical use of gastric function tests. Scand J Gastroenterol 5(suppl 6):9–46, 1970

Berg NO, et al: How to approach the child suspected of malabsorption. Acta Pediatr Scand 67:403–411, 1978

Gray GM: Carbohydrate digestion and absorption. Role of the small intestine. N Engl J Med 292:1225–1230, 1975

Olsen WA: A practical approach to diagnosis of disorders of intestinal absorption. N Engl J Med 285:1358–1361, 1971

Russell RI, Lee FD: Tests of small-intestinal function—digestion, absorption, secretion. Clin Gastroenterol 7:277–315, 1978

Thompson JC, Reeder DD, Bunchman HH: Clinical role of serum gastrin measurements in the Zollinger-Ellison syndrome. Am J Surg 124:250–261, 1972

Wormsley KG: Tests of pancreatic secretion. Clin Gastroenterol 7: 529–544, 1978

Text (ASCP)

Council on Clinical Chemistry
Batsakis JG, Briere RO, Markel SF: Diagnostic Enzymology. Chicago: American Society of Clinical Pathologists, 1972 (Cat. no. 45-2-026-00—out of print)

Audiovisual Aids (ASCP)

Trainer TD, Picoff RC, Duffell DR: Intestinal Malabsorption I: The Digestion and Absorption of Carbohydrates. Chicago: American Society of Clinical Pathologists, 1972 (Cat. no. 21-9-005-00—out of print)

Trainer TD, Picoff RC, Duffell DR: Intestinal Malabsorption II: Lactase Deficiency and Lactose Tolerance Test. Chicago: American Society of Clinical Pathologists, 1972 (Cat. no. 21-9-006-00—out of print)

Trainer TD, Picoff RC, Duffell DR: Intestinal Malabsorption III: D-Xylose Absorption Test. Chicago: American Society of Clinical Pathologists, 1972 (Cat. no. 21-9-007-00—out of print)

Trainer TD, Picoff RC, Duffell DR: Intestinal Malabsorption IV: Fats—Digestion and Absorption. Chicago: American Society of Clinical Pathologists, 1972 (Cat. no. 21-9-008-00—out of print)

Trainer TD, Picoff RC, Duffell DR: Intestinal Malabsorption V: The Qualitative Tests for Steatorrhea. Chicago: American Society of Clinical Pathologists, 1972 (Cat. no. 21-9-009-00—out of print)

Trainer TD, Picoff RC, Duffell DR: Intestinal Malabsorption VI: The Quantitative Tests for Steatorrhea. Chicago: American Society of Clinical Pathologists, 1972 (Cat. no. 21-9-010-00—out of print)

Trainer TD, Picoff RC, Duffell DR: Intestinal Malabsorption VII: Urinary Indican. Chicago: American Society of Clinical Pathologists, 1972 (Cat. no. 21-9-011-00—out of print)

Trainer TD, Picoff RC, Duffell DR: Intestinal Malabsorption VIII: Serum Carotene. Chicago: American Society of Clinical Pathologists, 1972 (Cat. no. 21-9-012-00—out of print)

Trainer TD, Picoff RC, Duffell DR: Intestinal Malabsorption IX: Small Bowel Biopsy. Chicago: American Society of Clinical Pathologists, 1972 (Cat. no. 21-9-013-00—out of print)

LEVEL II

Content

Pathophysiology—correlation of structure and function in gastrointestinal dysfunctional states

Diagnosis and management—specific approaches to the diagnosis of carbohydrate, protein, and other dietary malabsorptive states; discriminant function and multivariate analysis of test results for development of appropriate test strategies and for interpretation of test results in acid-peptic disease, malabsorption, and pancreatic diseases; the role of laboratory tests in the management of gastrointestinal disease

Goals

A more in-depth knowledge of gastroenterology to diagnose specific malabsorptive states, and to evaluate tests used in malabsorption and acid-peptic disease

Learning Aids

Bibliography

Alpers DH, Seetharam B: Pathophysiology of diseases involving intestinal brush-border proteins. N Engl J Med 296:1047-1050, 1977

Gardner JD: Receptors for gastrointestinal hormones. Gastroenterology 76:202-214, 1979

Pancreatic diarrheal syndromes—Medical Staff Conference, University of California, San Francisco. West J Med 123:290-296, 1975

Pearse AGE, Polak JM, Bloom SR: The newer gut hormones. Cellular sources, physiology, pathology and clinical aspects. Gastroenterology 72:746-761, 1977

Wingate D: The eupeptide system. A general theory of gastrointestinal hormones. Lancet 1:529-532, 1976

IMMUNOLOGIC DISORDERS

LEVEL I

Content

Pathophysiology—deficiencies and abnormalities of the immune response

Diagnosis—differential diagnosis of polyclonal gammopathy; strategies for detection of immunologic deficiency defects and for investigation of discrete hypergammaglobulinemia

Management—value of immunoglobulin determinations in the management of polyclonal and monoclonal gammopathies

Goals

Sufficient knowledge of the humoral immune system to test for immune deficiency; to conduct strategies for investigation of discrete hypergammaglobulinemia; and to interpret immunodiffusion and immunoelectrophoretic test results

Learning Aids

Bibliography

Conklin R, Alexanian R: Clinical classification of plasma cell myeloma. Arch Intern Med 134:139–143, 1975

Delaney WE: Identification and quantitation of immunoglobulins. Ann Clin Lab Sci 2:75–92, 1972

Hobbs JR: Monitoring myelomatosis. Arch Intern Med 135:125–130, 1975

Kyle RA: Multiple myeloma: Review of 869 cases. Mayo Clin Proc 50: 29–40, 1975

Martin NH: The immunoglobulins: A review. J Clin Pathol 22: 117–131, 1969

Moore EC, Meuwissen HJ: Immunologic deficiency disease. Approach to diagnosis. NY State J Med 73:2437–2445, 1973

Norman PS: The clinical significance of IGE. Hospital Practice 10: 41–49, 1975

Perry MC, Kyle RA: The clinical significance of Bence Jones proteinuria. Mayo Clin Proc 50:234–238, 1975

Ritzmann SE, Daniels JC: Serum Protein Abnormalities: Diagnostic and Clinical Aspects. Boston: Little Brown & Co, 1975

Roitt I: Essential Immunology, ed 3. Philadelphia: JB Lippincott Co, 1977

Schwartz RS: Therapeutic strategy in clinical immunology. N Engl J Med 280:367–374, 1969

Texts and Atlases (ASCP)

Council on Clinical Chemistry

Cawley LP: Electrophoresis and Immunoelectrophoresis. Chicago: American Society of Clinical Pathologists, 1966 (Cat. no. 45-2-009-00—out of print)

Cawley LP, et al: Basic Electrophoresis, Immunoelectrophoresis and Immunochemistry. Chicago: American Society of Clinical Pathologists, 1972 (Cat. no. 45-2-030-00—out of print)

Cawley LP, Minard BJ, Penn GM: Electrophoresis and Immunochemical Reactions in Gels: Techniques and Interpretation. Chicago: American Society of Clinical Pathologists, 1978 (Cat. no. 45-2-035-00)

Other Councils

Penn GM, Batya J: Interpretation of Immunoelectrophoretic Patterns. Chicago: American Society of Clinical Pathologists, 1978 (Atlas and slides: Cat. no. 15-A-001-00/Atlas only: Cat. no. 16-A-001-00)

Penn GM, Davis T: Identification of Myeloma Proteins. Chicago: American Society of Clinical Pathologists, 1975 (Cat. no. 45-A-002-00)

Audiovisual Aids (ASCP)

Penn GM: Clinical Pathologic Observations of Myeloproteins. Audiotape 42. Chicago: American Society of Clinical Pathologists, 1974 (Cat. no. 18-0-042-06—out of print)

Tucker ES III: Basic Concepts of Immunology. Chicago: American Society of Clinical Pathologists, 1978 (Cat. no. 47-A-034-00)

LEVEL II

Content

Pathophysiology—mechanisms of immunologic injury: correlation of structure and function; autoimmune disease; diseases caused by complex immunologic phenomena; organ transplantation

Diagnosis—test strategies for diagnosis of autoimmune diseases and diseases caused by complex immunologic phenomena; specific diagnosis of immune deficiency defects

Goals

Thorough understanding of the underlying mechanisms of immunologic injury; knowledge of autoimmune diseases; and the ability to develop test strategies for the diagnosis of autoimmune phenomena and for the specific diagnosis of immune deficiency defects

Learning Aids

Bibliography

Cohen AS: Laboratory Diagnostic Procedures in the Rheumatic Diseases, ed 2. Boston: Little Brown & Co, 1975

McCluskey RT, Hall CL, Colvin RB: Immune complex mediated diseases. Hum Pathol 9:71–84, 1978

Nakamura RM: Immunopathology: Clinical Laboratory Concepts and Methods. Boston: Little Brown & Co, 1974

Sell S: Introduction: Immune mechanisms in human disease. Hum Pathol 9:23–24, 1978

Tissue antigens and disease, editorial. Lancet 2:536–537, 1975

Tomasi TB: Secretory immunoglobulins. N Engl J Med 287:500–506, 1972

Texts and Atlas (ASCP)

Deodhar S, Nakamura RM: Atlas of Autoimmune Diseases. Chicago: American Society of Clinical Pathologists, 1976 (Cat. no. 15-A-002-00)

Nakamura RM, Deodhar S: Laboratory Tests in the Diagnosis of Autoimmune Disorders. Chicago: American Society of Clinical Pathologists, 1975 (Cat. no. 45-A-003-00)

Nakamura RM, et al: Autoantibodies to Nuclear Antigens (ANA): Immunochemical Specificities and Significance in Systemic Rheumatic Diseases. Chicago: American Society of Clinical Pathologists, 1978 (Cat. no. 45-A-004-00)

Audiovisual Aids (ASCP)

Nakamura RM: Diagnostic Evaluation of Autoimmune Diseases. Chicago: American Society of Clinical Pathologists, 1979 (Cat. no. 47-A-040-00)

Nakamura RM: Mechanisms of Autoimmune Disease. Chicago: American Society of Clinical Pathologists, 1979 (Cat. no. 47-A-039-00)

Tucker ES III: Mechanisms of Immunologic Tissue Injury. Chicago: American Society of Clinical Pathologists, 1978 (Cat. no. 47-A-035-00)

METABOLIC DISEASES

LEVEL I

Content

Pathophysiology—purine metabolism; porphyrin metabolism

Diagnosis—strategy for the diagnosis of acute gout; hyperuricemia and hypouricemia; qualitative testing for the porphyrias

Management—the role of uric acid determinations in the management of hyperuricemia

Goals

Comprehension of purine and porphyrin metabolism and their disorders, and familiarity with qualitative tests for porphyrias; the ability to diagnose acute gout; and to make differential diagnoses of hyperuricemic and hypouricemic states

Learning Aids

Bibliography

Bendersky G: Etiology of hyperuricemia. Ann Clin Lab Sci 5: 456–467, 1975

Elder GH, Gray CH, Nicholson DC: The porphyrias: A review. J Clin Pathol 25:1013–1033, 1972

Fessel WJ, Siegelaub AB, Johnson ES: Correlates and consequences of asymptomatic hyperuricemia. Arch Intern Med 132:44–54, 1973

Kelley WN: Current therapy of gout and hyperuricemia. Hospital Practice 11:69–76, 1976

Levere RD, Kappas A: The porphyric diseases of man. Hospital Practice 5:61–73, 1970

Paulus HE, et al: Clinical significance of hyperuricemia in routinely screened hospitalized men. JAMA 211:277–281, 1970

Pochedly C: Hyperuricemia in leukemia and lymphoma. Postgrad Med 55:93–99, 1974

Ramsdell CM, Kelley WN: The clinical significance of hypouricemia. Ann Intern Med 78:239–241, 1973

Rastegar A, Thier SO: The physiologic approach to hyperuricemia. N Engl J Med 286:470–476, 1972

With TK: Clinical porphyrin analyses: Indications and interpretations. Scand J Clin Lab Invest 38:501–505, 1978

Texts (ASCP)

Council on Clinical Chemistry

Bittner DL, Gambino SR: Uric Acid Assays. Historical, Clinical and Current Concepts. Chicago: American Society of Clinical Pathologists, 1970 (Cat. no. 45-2-025-00—out of print)

Patterson JN, Catanzaro C: The Porphyrins and the Porphyrias. Chicago: American Society of Clinical Pathologists, 1966 (Cat. no. 45-2-012-00—out of print)

LEVEL II

Content

Pathophysiology—biochemical defects in the porphyrias

Diagnosis—strategies for quantitative testing of the porphyrias

New developments in the biochemistry, diagnosis, and management of metabolic diseases

Goals

Thorough understanding of porphyrin metabolism and methods for the specific diagnosis of the porphyrias; familiarity with new developments in the biochemistry, diagnosis, and management of metabolic diseases

Learning Aids

Bibliography

Brodie MJ, Moore MR, Goldberg A: Enzyme abnormalities in the porphyrias. Lancet 2:699–701, 1977

Differential diagnosis of the hepatic porphyrias, editorial. Br Med J 4: 725–726, 1975

Kelley WN, Grobner W, Holmes E: Current concepts in the pathogenesis of hyperuricemia. Metabolism 22:939–959, 1973

Liang MH, Fries JF: Asymptomatic hyperuricemia: The case for conservative management. Ann Intern Med 88:666–670, 1978

Tschudy DP, Valsamis M, Magnussen CR: Acute intermittent porphyria: Clinical and selected research aspects. Ann Intern Med 83: 851–864, 1975

NUTRITIONAL DISORDERS

LEVEL II

Content

Pathophysiology—nutritional disturbance and vitamin deficiency states

Diagnosis—tests for the detection and diagnosis of nutritional and vitamin deficiency disorders

Management—the role of the laboratory in the management of nutritional disease

Goals

Knowledge of the pathophysiology of protein-calorie malnutrition and vitamin deficiency; familiarity with the role of the laboratory in the detection, diagnosis, and management of nutritional diseases

Learning Aids

Bibliography

Carlmark B, Reizenstein P: Comparison of methods to diagnose deficiency or malabsorption of vitamin B12. Scand J Gastroenterol 9 (suppl 29):39–42, 1974

DeLuca HF: Recent advances in our understanding of the vitamin D endocrine system. J Lab Clin Med 87:7–26, 1976

Herbert V: The five possible causes of all nutrient deficiency: Illustrated by deficiencies of vitamin B 12 and folic acid. Am J Clin Nutr 26:77–86, 1973

Laboratory tests in protein-calorie malnutrition. Lancet 1:1041–1042, 1973

Rosenberg IH: Folate absorption and malabsorption. N Engl J Med 293:1303–1308, 1975

Toskes P, Deren JJ: Vitamin B12 absorption and malabsorption. Gastroenterology 65:662–683, 1973

BIOCHEMISTRY OF NEOPLASIA

LEVEL I

Content

Pathophysiology—biochemical correlates, biochemical convergence, inappropriate cell products, hormone dependency, immunologic aspects

Diagnosis—the role of clinical enzymology in the detection and diagnosis of malignancies, acid and alkaline phosphatase, Regan isoenzyme; fetal proteins, alpha-fetoprotein and carcinoembryonic antigen (CEA) hormones, and ectopic hormone production

Management—the value of enzymes, fetal proteins, and hormones in the management of neoplastic disease

Goals

A basic knowledge of the biochemical correlates of neoplasia sufficient to understand the limited role of routine tests in the diagnosis and management of neoplastic diseases. the role of clinical enzymology in detection and diagnosis, and the use of fetal proteins and hormones in diagnosis and management

Learning Aids

Bibliography

Gittes R: Acid phosphatase reappraised. N Engl J Med 297: 1398–1399, 1977

Hansen HJ, et al: Carcinoembryonic antigen (CEA) assay. A laboratory adjunct in the diagnosis and management of cancer. Hum Pathol 5:139–147, 1974

Lange PH, et al: Alpha-fetoprotein and human chorionic gonadotropin in the management of testicular tumors. J Urol 118:593–596, 1977

McGuire WL, et al: Current status of estrogen and progesterone receptors in breast cancer. Cancer 39(suppl 6): 2934–2947, 1977

Nakamura RN, Plow EF, Edgington TS: Current status of carcinoembryonic antigen (CEA) and CEA-S assays in the evaluation of neoplasms of the gastrointestinal tract. Ann Clin Lab Sci 8:4–10, 1978

Rosen SW, et al: Placental proteins and their subunits as tumor markers. Ann Intern Med 82:71–83, 1975

Schwartz MK: Tumor markers, enzymes and hormonal markers. Clin Gastroenterol 5:653–663, 1976

Sugarbaker PH, Zamcheck N, Moore FD: Assessment of serial carcinoembryonic antigen (CEA) assays in postoperative detection of recurrent colorectal cancer. Cancer 38:2310–2316, 1976

Townsend RM: Enzyme tests and diseases of the prostate. Ann Clin Lab Sci 7:254–261, 1977

Welbourn RB: Current status of the apudomas. Ann Surg 185:1–12, 1977

Zamcheck N: The present status of carcinoembryonic antigen (CEA)

in diagnosis, detection of recurrence, prognosis and evaluation of therapy of colonic and pancreatic cancer. Clin Gastroenterol 5: 625–638, 1976

Zondag HA, Klein F: Clinical applications of lactate dehydrogenase isoenzymes: Alterations in malignancy. Ann NY Acad Sci 151: 578–586, 1968

Text (ASCP)

Council on Clinical Chemistry

Batsakis JG, Briere RO, Markel SF: Diagnostic Enzymology. Chicago: American Society of Clinical Pathologists, 1972 (Cat. no. 45-2-026-00 – out of print)

LEVEL II

Content

Pathophysiology – oncogenesis: genetic and immunologic aspects

Diagnosis – critical analysis of the value of laboratory tests in the detection and diagnosis of neoplasia; discriminant function analysis and multivariate analysis of test results

New developments in genetic, viral, immunologic, and environmental aspects of neoplasia. New methods of detection and diagnosis

Goals

In-depth knowledge of the genetic, viral, immunologic, and environmental aspects of neoplasia; and of the new methods of detection and diagnosis. Familiarity with theories of oncogenesis; and the ability to evaluate laboratory tests used for the detection and diagnosis of neoplasia.

Learning Aids

Bibliography

Broder S, Waldmann TA: The suppressor-cell network in cancer. N Engl J Med 299:1281–1284, 1335–1341, 1978

Cox RP, Ghosh NK: Current concepts in the ectopic production of fetal proteins and hormones by neoplastic cells. Am J Med Sci 275: 232–248, 1978

Drew SI: Immunological surveillance against neoplasia: An immunological quandary. Hum Pathol 10:5–14, 1979

Neville AM, Cooper EH: Biochemical monitoring of cancer. A review. Ann Clin Biochem 13:283–305, 1976

Tischler AS, et al: Neuroendocrine neoplasms and their cells of origin. N Engl J Med 296:919–925, 1977

Weber G: Enzymology of cancer cells. N Engl J Med 296:486–493, 541–551, 1977

INBORN ERRORS OF METABOLISM

LEVEL II

Content

Pathophysiology—genetically determined biologic variation, protein polymorphism, common inherited metabolic disorders

Diagnosis—neonatal screening, investigation of the symptomatic infant or child; screening of the apparently healthy population for heterozygotes; investigation of high-risk groups

Management—genetic counseling

Goals

Sufficient knowledge of genetics to understand the mechanisms of inherited metabolic disease, screening and case-finding techniques for common inborn errors of metabolism, and the basis of genetic counseling

Learning Aids

Bibliography

Bradley GM: Urinary screening tests in the infant and young child. Hum Pathol 2:309–320, 1971

Dubowitz V: Screening for Duchenne muscular dystrophy. Arch Dis Child 51:249-251, 1976

Erbe RW: Principles of medical genetics. N Engl J Med 294:318, 480-482, 1976

Guthrie R: Mass screening for genetic disease. Hospital Practice 7: 93-100, 1972

Harris H: The Principles of Human Biochemical Genetics, ed 2. New York: Elsevier North-Holland Inc, 1975

Neufeld EF: Mucopolysaccharidoses: The biochemical approach. Hospital Practice 7:107-113, 1972

Raine DN, et al: Screening for inherited metabolic disease by plasma chromatography (Scriver) in a large city. Br Med J 3:7-13, 1972

Rennert OM: Enzymatic screening of genetic diseases. Ann Clin Lab Sci 7:262-268, 1977

Singer JD, Cotlier E, Krimmer R: Hexosaminidase A in tears and saliva for rapid identification of Tay-Sachs disease and its carriers. Lancet 2:1116-1119, 1973

Stanbury JB, Wyngaarden JB, Frederickson DS: The Metabolic Basis of Inherited Disease, ed 3. New York: McGraw-Hill Book Co, 1972

Webb JR, et al: PKU screening—Is it worth it? Can Med Assoc J 108: 328-329, 1973

LEVEL III

Content

Pathophysiology—inborn errors of protein, carbohydrate, and lipid metabolism

Diagnosis—critical evaluation of screening programs for inborn errors of metabolism; prenatal diagnosis

New developments in the understanding, detection, and management of inborn errors of metabolism

Goals

Thorough understanding of inborn errors of metabolism and of the current status of prenatal diagnosis; and the ability to evaluate programs for screening of inborn errors of metabolism

Learning Aids

Bibliography

Davidson RG, Rattazzi MC: Prenatal diagnosis of genetic disorders: Trials and tribulations. Clin Chem 18:179–187, 1972

Gerbie AB, Simpson JL: Antenatal detection of genetic disorders. Postgrad Med 59:129–136, 1976

Milunsky A: Prenatal diagnosis for genetic disorders. N Engl J Med 295:377–380, 1976

Milunsky A, Alpert E: Prenatal diagnosis of neural tube defects. I. Problems and pitfalls: Analysis of 2495 cases using the alpha-feto protein results. Obstet Gynecol 48:6–12, 1976

Nadler HL: Prenatal diagnosis of inborn defects: A status report. Hospital Practice 10:41–51, 1975

Report of UK collaborative study on alpha-fetoprotein in relation to neural tube defects: Maternal serum-alpha-fetoprotein measurement in antenatal screening for anencephaly and spina bifida in early pregnancy. Lancet 1:1323–1338, 1977

THERAPEUTIC DRUG MONITORING AND TOXICOLOGY

LEVEL II

Content

Pharmacology and pathophysiology—basic pharmacokinetics; pharmacology of cardiac glycosides, anticonvulsants, antiarrhythmic agents, and antidepressants; pathophysiologic effects of salicylates, alcohol, barbiturates, carbon monoxide, antidepressants

Diagnosis – interpretation of toxicologic analyses

Management – interpretation of therapeutic drug levels and laboratory management of poisoning

Goals

Comprehension of the basic pharmacology of cardiac glycosides, anticonvulsants, antiarrhythmic agents, and common poisons and drugs; of the pathology of drug-overdose; and of fundamental pharmacokinetics. The ability to interpret and provide clinical advice for common toxicologic and drug analyses.

Learning Aids

Bibliography

Therapeutic Drug Monitoring

Atkinson AJ: Individualization of anticonvulsant therapy. Med Clin North Am 58:1037–1049, 1974

Doherty JE: How and when to use the digitalis serum levels. JAMA 239:2594–2596, 1978

Dvorchik BH, Vesell ES: Pharmacokinetic interpretation of data gathered during therapeutic drug monitoring. Clin Chem 22:868–878, 1976

Frings CS: Drug screening. CRC Crit Rev Clin Lab Sci 4:357–382, 1973

Greenblatt DJ, Koch-Weser J: Clinical pharmacokinetics. N Engl J Med 293:702–705, 293:964–970, 1975

Koch-Weser J, Duhme DW, Greenblatt DJ: Influence of serum digoxin concentration measurements on frequency of digitoxicity. Clin Pharmacol Ther 16:284–287, 1974

Leal KW, Troupin AS: Clinical pharmacology of anti-epileptic drugs: A summary of current information. Clin Chem 23:1964–1968, 1977

Luchins D, Ananth J: Therapeutic implications of tricyclic antidepressant plasma levels. J Nerv Ment Dis 162:430–436, 1976

Shapiro B, Kollmann GJ, Heine WI: Pitfalls in the application of digoxin determinations. Semin Nucl Med 5:205–220, 1975

Sheiner LB, et al: Improved computer-assisted digoxin therapy: A method using feedback measured serum digoxin concentrations. Ann Intern Med 82:619–627, 1975

Smith TW, Haber E: Digoxin intoxication: The relationship of clinical presentation to serum digoxin concentration. J Clin Invest 49: 2377–2386, 1970

Winter PN, Miller JN: Carbon monoxide poisoning. JAMA 236: 1502–1504, 1976

Toxicology

Finkle BS: Forensic toxicology of drug abuse: A status report. Anal Chem 44:18A–31A, 1972

Lundberg GD, Walberg CG, Pantlik VA: Frequency of clinical toxicology test-ordering (primarily overdose cases) and results in a large urban general hospital. Clin Chem 20:121–125, 1974

McCarron MM, Walberg CB, Lundberg GD: Are emergency toxicology measurements really used? Clin Chem 20:116–120, 1974

Winek CL: Tabulation of therapeutic, toxic, and lethal concentrations of drugs and chemicals in blood. Clin Chem 22:832–836, 1976

Text (ASCP)

Council on Clinical Chemistry

Bucklin RV, et al: I. Toxicology for the medical examiner, in Fuller JB (ed): Selected Topics in Clinical Chemistry II. Chicago: American Society of Clinical Pathologists, 1975, pp 1–24 (Cat. no. 45-2-033-00—out of print)

Audiovisual Aid (ASCP)

Finley PR: Digoxin Assay: Laboratory and Clinical Aspects. Chicago: American Society of Clinical Pathologists, 1978 (Cat. no. 47-8-027-00)

Miscellaneous Learning Aids (ASCP)

Froede RC: Patterns of drug abuse, in Baer DM (ed): Technical Improvement Service. Number 19. Shilling JM (section ed). Chicago: American Society of Clinical Pathologists, 1974, pp 7–25 (Cat. no. 63-0-019-00—out of print)

Sobota JT: Toxicology screening tests, in Baer DM (ed): Technical Improvement Service. Number 21. Chicago: American Society of Clinical Pathologists, 1975, pp 72–91 (Cat. no. 63-0-021-00–out of print)

LEVEL III

Content

Chemistry and pathophysiology—industrial and occupational hazards, heavy metal poisons; organic solvents; pesticides and herbicides; miscellaneous drugs and poisons

Diagnosis and management—chemical investigation of poisonous materials, including therapeutic drugs, illicit drugs, industrial, and occupational hazards; new developments in therapeutic monitoring

Goals

A broader comprehension of the chemistry and pathophysiology of industrial and occupational hazards; and of advanced therapeutic monitoring; the ability to evaluate a wide variety of toxicologic problems and to make specific diagnoses

Learning Aids

Bibliography

Amdisen A: Monitoring of lithium treatment through determination of lithium concentration. Dan Med Bull 22:277–291, 1975

Biggs JT: Clinical pharmacology and toxicology of antidepressants. Hospital Practice 13:79–84, 1978

Broughton A, Strong JE: Radioimmunoassay of antibiotics and chemotherapeutic agents. Clin Chem 22:726–732, 1976

Burch RE, Sullivan JF: Diagnosis of zinc, copper, and manganese abnormalities in man. Med Clin North Am 60:655–660, 1976

Chisolm JJ Jr: The continuing hazard of lead poisoning. Hospital Practice 8:127–135, 1973

Frings CS: The role of the laboratory in effective therapeutic monitoring of lithium, digoxin, procainamide, phenobarbital and diphenylhydantoin. Lab Med 6:31–33, 1975

Garry VF Jr, Weston JT: Environmental pathology: New directions. Hum Pathol 10:1–3, 1979

Greenblatt DJ, et al: Pharmacokinetic approach to the clinical use of lidocaine intravenously. JAMA 236:273–277, 1976

Houk VN, et al: Increased lead absorption and lead poisoning in young children. A statement by the Center for Disease Control. J Pediatr 87:824–830, 1975

Jenne JW, et al: Pharmacokinetics of theophylline. Application to adjustment of the clinical dose of aminophylline. Clin Pharmacol Ther 13:349–360, 1972

Martin HF: The role of the clinical laboratory in assessing occupationally related diseases. Hum Pathol 9:487–491, 1978

McHenry LE: Toxicology in a community hospital. Lab Med 6: 15–23, 1975

O'Dell BL: Biochemistry of copper. Med Clin North Am 60:687–703, 1976

Spears AB, et al: Serum-zinc as index of zinc status. Lancet 2:526, 1976

Sunderman FW Jr: Current status of zinc deficiency in the pathogenesis of neurological, dermatological and musculoskeletal disorders. Ann Clin Lab Sci 5:132–145, 1975

Werner M, Sutherland EW, Abramson FP: Concepts for the rational selection of assays to be used in monitoring therapeutic drugs. Clin Chem 21:1368–1371, 1975

Section 3

DIAGNOSTIC APPLICATION OF LABORATORY DATA

LEVEL I

Content

Normal values—determination of the normal range, parametric vs nonparametric methods

Reference values

Clinical decisions—detection, exclusion, confirmation, assessment of change

Diagnostic validity of laboratory tests—Bayesian principles, sensitivity, specificity, predictive value, efficiency

Biochemical profiling—interpretation, qualitative approaches

Construction of interpretative and diagnostic search algorithms

Goals

The ability to determine normal and reference values and to interpret routine biochemical profiles; familiarity with the varieties of clinical decisions that demand laboratory tests; knowledge of the principles of diagnostic validity, Bayes' theorem; and of the elements of elaborating diagnostic test strategies

Learning Aids

Bibliography

Normal Values

Elveback LR, Guillier CL, Keating FR: Health, normality, and the ghost of Gauss. JAMA 211:69–75, 1970

Mainland D: Remarks on clinical "norms." Clin Chem 17:267–274, 1971

Martin HR, Gudzinowica BJ, Fanger H: Normal Values in Clinical Chemistry. Definition and Use. New York: Marcel Dekker Inc, 1975

Reed AH, Henry RJ, Mason WB: Influence of statistical method used on the resulting estimate of normal range. Clin Chem 17:275–284, 1975

Williams GZ, et al: Biological and analytic components of variation in long-term studies of serum constituents in normal subjects. I. Objectives, subject selection, laboratory procedures, and estimation of analytic deviation; II. Estimating biological components of variations; III. Physiological and medical implications. Clin Chem 16: 1016–1021, 122–1027, 1028–1032, 1970

Reference Values

Alström T, et al: Recommendations concerning the collection of reference values in clinical chemistry and activity report. Scand J Clin Lab Invest 35(suppl 144):1–44, 1975

Dybkaer R, et al: Statistical terminology in clinical chemistry reference values. Scand J Clin Lab Invest 35(suppl 144):45–74, 1975

Gräsbeck R: Terminology and the biological aspects of reference values, in Benson ES, Rubin M (eds): Logic and Economics of Clinical Laboratory Use. New York: Elsevier North-Holland Inc, 1978

Sunderman FW Jr: Current concepts of "normal values," "reference values," and "discrimination values" in clinical chemistry. Clin Chem 21:1873–1877, 1975

Clinical Decisions

Burke MD: Diagnosis: Art or Science? Diagnostic Medicine 1:25–32, 1978

Campbell EJ: Basic science, science and medical education. Lancet 1:134–136, 1976

Feinstein AR: Clinical Judgment. Baltimore: Williams & Wilkins Co, 1967

Kassirer JP, Gorry GA: Clinical problem solving: A behavioral analysis. Ann Intern Med 89:245–255, 1978

Lusted LB: Introduction to Medical Decision Making. Springfield, Ill: Charles C Thomas Publisher, 1968

Murphy EA: The Logic of Medicine. Baltimore: Johns Hopkins University Press, 1976

Diagnostic Validity

Galen RS, Gambino SR: Beyond Normality: The Predictive Value and Efficiency of Medical Diagnosis. New York: John Wiley & Sons Inc, 1975

McNeil BJ, Keeler E, Adelstein SJ: Primer on certain elements of medical decision making. N Engl J Med 293:211–215, 1975

Biochemical Profiling

Reece RL, Hobbie RK: Computer evaluation of chemistry values: A reporting and diagnostic aid. Am J Clin Pathol 57:664–675, 1972

Ward PC: Chemical profiles of disease. Hum Pathol 4:47–65, 1973

Whitehead TP, Wooton ID: Biochemical profiles for hospital patients. Lancet 2:1439–1443, 1974

Test Strategies

Burke MD: Clinical decision making: The role of the laboratory, in Benson ES, Rubin M (eds): Logic and Economics of Clinical Laboratory Use. New York: Elsevier North-Holland Inc, 1978, p 59

Crosby WH: Chess and combat: The algorithm in medicine. JAMA 238:2721, 1977

Feinstein AR: An analysis of diagnostic reasoning. 3. The construction of clinical algorithms. Yale J Biol Med 47:5–32, 1974

Knuth DE: Algorithms. Sci Am 236:63–80, 1977

Texts (ASCP)

Council on Clinical Chemistry

Barnett RN (ed): Statistical Methods in the Clinical Laboratory. Chicago: American Society of Clinical Pathologists, 1968 (Cat. no. 45-2-016-00—out of print)

Other Councils

Mukherjee KL: Introductory Mathematics for the Clinical Laboratory. Chicago: American Society of Clinical Pathologists, 1979 (Cat. no. 45-9-006-00)

Tischer RG, Turner BH, Hogan SC: Problem Solving in Medical Technology and Microbiology. Chicago: American Society of Clinical Pathologists, 1979 (Cat. no. 45-7-010-00)

Audiovisual Aid (ASCP)

Lundberg G Jr: Patient Focused Lab. Audiotape 33. Chicago: American Society of Clinical Pathologists, 1971 (Cat. no. 18-0-033-06—out of print)

LEVEL II

Content

Cost-effectiveness analysis

Cost-benefit analysis

Screening, case-finding, biochemical profiling

Clinical decision analysis

Goals

The ability to comprehend and conduct quantitative approaches to cost-benefit and cost-effectiveness analysis; clinical decision analysis; and critical analysis of screening, biochemical profiling, and case-finding techniques; and the ability to construct clinical algorithms and decision trees

Learning Aids

Bibliography

Cost-effectiveness

Ayres JC: Decelerating skyrocketing health-care cost. N Engl J Med 296:391–393, 1977

Cost-effectiveness studies, editorial. Br Med J 2:848–849, 1978

Griner PF: Use of laboratory tests in a teaching hospital: Long term trends. Reductions in use and relative cost. Ann Intern Med 90: 243–248, 1979

Weinstein MC, Stason WB: Foundations of cost-effectiveness analysis for health and medical practices. N Engl J Med 296:716–721, 1977

Cost-benefit

McNeil BJ, et al: Measures of clinical efficacy. Cost-effectiveness calculations in the diagnosis and treatment of hypertensive renovascular disease. N Engl J Med 293:216–221, 1975

Pauker SG, Kassirer JP: Therapeutic decision making: A cost-benefit analysis. N Engl J Med 293:229–234, 1975

Screening, Case Finding, and Profiling

Ahlvin RC: Biochemical screening—a critique. N Engl J Med 283: 1084–1086, 1970

Bates B, Yellin JA: The yield of multiphasic screening. JAMA 222: 74–78, 1972

Bradwell AR, Carmalt MHB, Whitehead TP: Explaining the unexpected abnormal results of biochemical profile investigations. Lancet 2:1071–1074, 1974

Durbridge TC, et al: Evaluation of benefits of screening tests done immediately on admission to hospital. Clin Chem 22:968–971, 1976

Holland WW: Taking stock. Lancet 2:1494–1497, 1974

Knox EG: Multiphasic screening. Lancet 2:1434–1436, 1974

Sackett DL: The usefulness of laboratory tests in health-screening programs. Clin Chem 19:366–372, 1973

Sharp CLEH, Keen H: Presymptomatic Detection and Early Diagnosis. A Critical Appraisal. London: Pitman Publishing Ltd, 1968

Whitehead TP, Wootton IDP: Biochemical profiles for hospital patients. Lancet 2:1439–1443, 1974

Clinical Decision Analysis

Ingelfinger FJ: Decision in medicine. N Engl J Med 293:254–255, 1975

Pauker SG: Coronary artery surgery: The use of decision analysis. Ann Intern Med 85:8–18, 1976

Ransohoff DF, Feinstein AR: Is decision analysis useful in clinical medicine? Yale J Biol Med 49:165–168, 1976

Schwartz WB, et al: Decision analysis and clinical judgment. Am J Med 55:459–472, 1973

Sisson JC, Schoomaker EB, Ross JC: Clinical decision analysis. The hazard of using additional data. JAMA 236:1259–1263, 1976

LEVEL III

Content

Actuarial and process tracing approaches in problem-solving research

New approaches to the application and interpretation of laboratory tests—multivariate analysis of test results; computerized diagnostic systems

Goals

Advanced knowledge of research in human problem solving with particular attention to medical problem solving; advanced knowledge in multivariate analytic techniques with computer applications

Learning Aids

Bibliography

Elstein AS, Shulman LS, Sprafka SA: Medical Problem Solving: An Analysis of Clinical Reasoning. Cambridge, Mass: Harvard University Press, 1978

Fisher RA: The use of multiple measurements in taxonomic problems. Annals of Eugenics 7:179–188, 1936

Grams RR, Johnson EA, Benson ES: Laboratory data analysis system: I. Introduction and overview; II. Analytic error limits; III. Multivariate normality; IV. Multivariate diagnosis; V. Trend analysis; VI. System summary. Am J Clin Pathol 58:177–219, 1972

Lachenbruck PA: Discriminant Analysis. New York: Hafner Press, 1975

Ladenson AH, Lewis JW, Boyd JC: Initial studies of test selection and pattern recognition in the differential diagnosis of hypercalcemia, in Benson ES, Rubin M (eds): Logic and Economics of Clinical Laboratory Use. New York: Elsevier North-Holland Inc, 1978, p 187

Newell A, Simon HA: Human Problem Solving. Englewood Cliffs, NJ: Prentice-Hall Inc, 1972

Pauker SG, et al: Towards the simulation of clinical cognition. Taking a present illness by computer. Am J Med 60:981–996, 1976

Winkel P: Patterns and clusters—multivariate approach for interpreting clinical chemistry results. Clin Chem 19:1329–1338, 1973

Section 4

LABORATORY ADMINISTRATION

Content

Fundamental principles of successful management—leadership requirements

Laboratory organization—policies, rules, regulations

Personnel management—interviewing laboratory personnel; personnel records; promotions, demotions, dismissals

Business methods—accounting, purchasing, and inventory

Inspection and accreditation

Laboratory programs—morale; efficiency; continuing education; safety; quality control; preventive maintenance

Goals

Comprehension of basic business methods and management principles to run an effective community hospital laboratory; familiarity with requirements for inspection and accreditation

Learning Aids

Bibliography

Bailey RM, Tierney TM Jr: Economic Perspectives on Clinical Laboratories. Berkeley, Calif: University of California Press, 1975

Bennington JL, Westlake GE, Louvau GE: Financial Management of the Clinical Laboratory. Baltimore: University Park Press, 1974

Fischer EF Jr, Sherrick JC: On the Sharing Centralization & Consolidation of Health Facility Laboratories & Related Diagnostic Services: An Annotated Bibliography. DHEW Publication No. (HRA) 74-3036. Washington, DC: US Government Printing Office, 1971

Moore RF (ed): AMA Management Handbook. New York: American Management Association, 1970

Newell JE: Laboratory Management. Boston: Little Brown & Co, 1972

Westlake GE, Bennington JL: Automation and Management in the Clinical Laboratory. Baltimore: University Park Press, 1971

Audiovisual Aids (ASCP)

Hoffman G: Programmable Desk Top Calculators. Audiotape 35. Chicago: American Society of Clinical Pathologists, 1971 (Cat. no. 18-0-035-06—out of print)

Monroe C: Personnel Problems. Audiotape 34. Chicago: American Society of Clinical Pathologists, 1971 (Cat. no. 18-0-034-06—out of print)

SUGGESTED REFERENCES

Beeler MF: Interpretations in Clinical Chemistry. Chicago: American Society of Clinical Pathologists, 1978

Beeson PB, McDermott W (eds): Textbook of Medicine, ed 14. Philadelphia: WB Saunders Co, 1975

Galen RS, Gambino SR: Beyond Normality: The Predictive Value and Efficiency of Medical Diagnosis. New York: John Wiley & Sons Inc, 1975

Henry JB (ed): Todd-Sanford-Davidsohn Clinical Diagnosis and Management by Laboratory Methods, ed 16. Philadelphia: WB Saunders Co, 1979

Henry RJ, Cannon DC, Winkleman JW (eds): Clinical Chemistry: Principles and Technics, ed 2. New York: Harper & Row Publishers Inc, 1974

Maxwell MH, Kleeman CR (eds): Clinical Disorders of Fluid and Electrolyte Metabolism, ed 2. New York: McGraw-Hill Book Co, 1972

Rose BD (ed): Clinical Physiology of Acid Base and Electrolyte Disorders. New York: McGraw-Hill Book Co, 1977

Schrier RW (ed): Renal and Electrolyte Disorders. Boston: Little Brown & Co, 1976

Sherlock S: Diseases of the Liver and Biliary System, ed 5. Philadelphia: FA Davis Co, 1975

Tietz N (ed): Fundamentals of Clinical Chemistry, ed 2. Philadelphia: WB Saunders Co, 1976

Whitby LG, Percy-Robb IW: Lecture Notes on Clinical Chemistry, Smith AF (ed). Philadelphia: JB Lippincott Co, 1975

Williams RH: Textbook of Endocrinology, ed 5. Philadelphia: WB Saunders Co, 1974

Zilva JF: Clinical Chemistry in Diagnosis and Treatment, ed 2. Chicago: Year Book Medical Publishers Inc, 1976